Astronautical Engineering

AF799627

Series Editor
Scott Madry

More information about this series at http://www.springer.com/series/5495

New Space Ventures

Ram S. Jakhu • Joseph N. Pelton
Yaw Otu Mankata Nyampong

Space Mining and Its Regulation

Springer

Published in association with
Praxis Publishing
Chichester, UK

PRAXIS

Ram S. Jakhu
Institute of Air and Space Law
McGill University
Montreal, QC, Canada

Joseph N. Pelton
Executive Board, International Association for the Advancement of Space Safety,
Arlington, VA, USA

Yaw Otu Mankata Nyampong
Pan African University
African Union Commission
Addis Ababa, Ethiopia

ISSN 2365-9599 ISSN 2365-9602 (electronic)
Springer Praxis Books
ISBN 978-3-319-81848-1 ISBN 978-3-319-39246-2 (eBook)
DOI 10.1007/978-3-319-39246-2

© Springer International Publishing Switzerland 2017
Softcover reprint of the hardcover 1st edition 2016
This work is subject to copyright. All rights are reserved by the Publisher, whether the whole or part of the material is concerned, specifically the rights of translation, reprinting, reuse of illustrations, recitation, broadcasting, reproduction on microfilms or in any other physical way, and transmission or information storage and retrieval, electronic adaptation, computer software, or by similar or dissimilar methodology now known or hereafter developed.
The use of general descriptive names, registered names, trademarks, service marks, etc. in this publication does not imply, even in the absence of a specific statement, that such names are exempt from the relevant protective laws and regulations and therefore free for general use.
The publisher, the authors and the editors are safe to assume that the advice and information in this book are believed to be true and accurate at the date of publication. Neither the publisher nor the authors or the editors give a warranty, express or implied, with respect to the material contained herein or for any errors or omissions that may have been made.

Cover illustration: Courtesy of Deep Space Industries.

Printed on acid-free paper

This Springer imprint is published by Springer Nature
The registered company is Springer International Publishing AG Switzerland

Foreword

Simply put, the authors of *Space Mining and Its Regulation* provide in this rather concise book a comprehensive view of the history, technical challenges, current status, and future probabilities of mining off-Earth space resources. This they do in breadth, depth, and in very accessible detail. They discuss the evolution toward space mining as a natural flow of the human use of space. As presented, the accurate history and analysis makes the author's views and presentations invaluable. It thus provides critical information both to those responsible for formulating space resource mining policies and those formulating and following laws designed and intended to implement the policies in a relatively global-interest context.

The book is thus an extraordinarily useful—perhaps even rare—resource, not just for the daily practitioner of space law relating to the mining of space resources but also to the broad spectrum of relevant policy makers, programmers, and project implementers. In many respects, the subject matter of the book is presented in an intriguingly simplistic as well as highly professional fashion that will attract an average interested and nontechnically oriented reader. The subject matters addressed specifically and in detail cover issues and positions that need to be taken regarding the present and long-term importance of off-Earth natural resources, i.e., exactly why they are needed for the current and future challenges of the world population.

Included in the discussions are the various transportation systems and locations for mining a variety of usable space resources; the use of robotic systems for mining purposes; the role of governmental activities and applicable policies in current and future resource mining; the involvement of private sector initiatives, principally by emerging US companies seeking to take advantage of the unfolding accessibility to a variety of different space and space-related resources; the past and current activities of the USA, the former Soviet Union, and now Russia in exploring and developing space resource mining capabilities, as well as those practices and planning pursuits in Japan, China, India, and also in Europe, Canada, and certain other nations located in the western hemisphere. All of these activities are explored by the authors in the context of the existing and potential international and various national regulatory environments. A close look is taken by the authors at existing and potential national space laws relating to resource exploitation.

Normally, daily or otherwise routine adoption of laws for domestic and even international purposes is simply the implementing of underlying policies already formulated and either existing or soon to come into effect. There has to be a reasonable understanding of the underlying philosophic construct driving all components of human space migration before the mining of space resources can be carried out in a responsible fashion. Otherwise, the sporadic and disparate mining of space resources will affect that migration in perhaps an irretrievably negative fashion. The mining and underlying costs must be for globally shared purposes and conducted with that overall aim or construct as a constant guiding principle. The cost effect of off-Earth mining may well be prohibitive without the broader objective of species migration and survival serving as a significant motivating factor. The book's authors have shown clearly, carefully, and unmistakably that the absence of pressingly definitive policies regarding the use of off-Earth space resources has forced the traditional law makers and space-related law practitioners to assume much responsibility normally left to the policy formulators. Bringing people together globally for a pressingly common, but little understood, objective is one of the greatest challenges historically faced by constantly evolving cultures, societies, and civilizations. The authors bring much of the global population together in a very pragmatic fashion to focus on the details of mining space resources for the benefit of humans living and working in space, as well as on Earth's surface.

The authors have identified a number of the different start-up, evolving, and established companies across the globe with programs or dedicated activities to explore usable space resources and exploit them financially to the greatest reasonable extent practicable. In addition to describing the flight potential objectives of these companies, the applicability of current space-related mining laws to these objectives is set forth. But some of the real issues emphasized by the authors with respect to identifying potentially usable space resources are the applicability of existing laws and the need to develop and fine-tune not only these laws and regulations but address the need for potentially unique laws as resource identification and usability unfold. In this context, and as relevant dependency on rapidly developing technologies necessary for space resource extraction and practical applications becomes clear and available, the need for an overall jurisprudence underlying resource extraction and applications, both on Earth and off, will become clear.

This book presents the underlying policies and available technologies, present and future, necessary for supporting migration off Earth. This is particularly clear in the context of learning how to use space resources, once they are mined successfully, for space habitation to support those mining efforts in situ. With respect to mining lunar resources, the authors are keenly aware of the need to know for what purposes the resources are being mined, and indeed have as thorough a knowledge as possible, for example, of the Moon's origin and composition, and whether and under what circumstances expanding mining undertakings might or will disturb its physical relationship with Earth and, indeed, orbiting habitats such as the current International Space Station (ISS). The flow of information set forth in the book is orderly, logical, and historically accurate, particularly with respect to the variety of space systems that have existed and currently exist for the retrieval, transportation, and delivery of mined space resources. Again, the book carefully sets forth the underlying policies and technologies necessary for migration off Earth and learning how to use space resources, once mined successfully, for space habitation and production of resources usable on Earth as well as in space.

In the context of space resources available, or becoming available, for commercial mining purposes, the authors address the rise of new companies resulting, often uniquely, by contracting and partnering with established companies, such as Northrop Grumman and the birth of commercial space industries over the past 15 or so years. These companies rely primarily on relatively modest start-up capital and bottom-up decision making using constantly evolving technology based on software development that is transforming the way space business is and will be conducted. In a revealing and fascinating fashion, the authors address the issues raised by these new companies and their management style, e.g., flatter and more flexible organizations that are consumer focused, innovative, and with owner-management willingness toward risk-taking. The companies by-and-large are organized in a way that focuses on new technology-oriented problem solving.

The authors document a zealous management impatience among all the new space mining companies with restrictive regulations. They are constantly pressing to move ahead rapidly. The authors focus primarily on the United States as the principal source of these new types of business formats and operational infrastructures that reflect activities emphasizing private sector pursuits in addressing space resources mining. They also present a marvelous hint of currently existing companies that could clearly identify and exploit usable resources offered by "potentially mineable" asteroids and Earth's moon.

Finally, the authors bravely tackle how they believe the national, international, and global communities will react to the developing manner of the evolving variety of interests, technologies, and methodologies for capturing and using space resources. In addition to presenting excellent and very readable histories of pursuits to capture and use space resources through often unique methodologies, the authors pose and address the final question for the moment, i.e., "How Is the World Likely to React to these Developments?" The book serves both as a fascinating read and particularly as a very usable text to which space lawyers and policy makers can easily resort for accurate and relevant information in furtherance of their respective professional responsibilities.

Dr. George S. Robinson

Preface

This book is designed as a "one-stop shopping" guide to the newly emerging field of space mining. The chapters that follow seek to provide a review of the past, current, and planned activities of various national space agencies of the world—as well as new commercial enterprises—in their relevant efforts to explore and exploit the resources of the Solar System. This review covers the Moon, Mercury, Venus, Mars, Jupiter and its Moon as well as asteroids and even comets. It includes an overview of the exploratory and scientific activities of the United States, USSR/Russia, Europe, Japan, China, India, Canada, and others in terms of space initiatives of national space agencies.

This book also provides a review of the current activities of the new space mining ventures, including Planetary Resources, Deep Space Industries, Moon Express, and Shackleton Energy. The book describes the international and national legal and regulatory frameworks (or the absence thereof) within which space mining is being and will be developed or undertaken. Special attention is paid to the legal issues related to existing and evolving international liabilities, property rights, and national licensing systems applicable to private entities aspiring to engage in space mining.

Clearly there are great technological, ecological, and legal and policy challenges to be met and resolved before space mining can make the transition from aspiration to reality. This book attempts to provide useful background as to past and current activities as well as to offer a guide to that future as well.

Montreal, QC, Canada — Ram S. Jakhu
Arlington, VA, USA — Joseph N. Pelton
Addis Ababa, Ethiopia — Yaw Otu Mankata Nyampong

Acknowledgements

This book is a significantly expanded and updated version of a study that was carried out in 2013 by Ram S. Jakhu and Yaw O. M. Nyampong for the Canadian Space Agency (CSA). The CSA, which holds copyright to the original study, has authorized and encouraged the authors to publish the study under their names. We are indebted to the CSA for its generosity and support.

Tanveer Ahmed helped us in the collection of data for some chapters of this book. We express our gratitude to Tanveer. Dr. Joseph Pelton, the former dean of the International Space University, has added a significant amount of new text with regard to the technical challenges that space mining faces in the years ahead and has participated fully in the complete rewriting of the original study and providing updated materials throughout.

However, as always and notwithstanding the above-mentioned invaluable help and support, the authors remain exclusively responsible for any errors contained in this book.

THE ARTHUR C. CLARKE FOUNDATION

The Arthur C. Clarke Foundation was established in 1983 in Washington, D.C., as part of World Communications Year celebrations at the United Nations, an international event sponsored by the United Nation's International Telecommunication Union (ITU). The Foundation was created to recognize and promote the extraordinary contributions of Arthur C. Clarke to the world and to promote the use of space and telecommunications technology for the benefit of humankind.

The Foundation is dedicated to enhancing Sir Arthur Clarke's legacy and to share that opportunity with like-minded institutions and especially sister organizations such as the Arthur C. Clarke Center for the Human Imagination and the Arthur C. Clarke Institute for Space Education.

The Foundation draws its inspiration from an individual whose range of creativity is unimaginably wide. At one extreme is Clarke the physicist who, at age 28, envisioned a time wherein geosynchronous platforms—extraterrestrial relays—could be used for global communication. At the other is Clarke as the most inspiring science fiction writer of his age, and his relentless and profound faith in humanity's ability to meet, even to elevate, its moral obligations to the planet upon which we live.

To the world, Arthur C. Clarke was a visionary, known not only for his science fiction novels such as Childhood's End, Rendezvous with Rama, and 2001: A Space Odyssey, but also for his scientific publications on space, energy, and the oceans. He is perhaps most famous for envisioning a global network of geosynchronous telecommunications satellite in 1945, as well as conceptualizing the "space elevator"—an elevator from Earth's surface to orbit, and ocean thermal energy conversion (OTEC).

The Arthur C. Clarke Foundation regards Sir Arthur's work as an unparalleled synthesis of science, literature, and social concern. History will list him among the few whose insights ranged most broadly in our comprehension of the universe we live in, the way we live in it, and the responsibility we have to improve our world. The scope of his vision is presented in the book *The Oracle of Colombo: How Arthur C. Clarke Revealed the Future*.

The Foundation exists to

- Stimulate creative use of communications technologies and social resources to improve health, education, and the quality of life for people everywhere, with emphasis on the needs of developing countries.
- Integrate science and technology with literature, film, and other means of outreach to enhance recognition of our increasingly complex, interconnected world.
- Deepen public understanding of science and technology and their impact on humanity and our world.

To carry out its mission, the Foundation has created and oversees annual awards, educational programs, video productions, lectures, fellowships, travel grants, and endowments to commemorate the life and works of Arthur C. Clarke who died in 2008.

GSI Global Space Institute

The Global Space Institute was formed in 2014 and has offices in the United States and Canada, but is organized and committed to support the outer space community worldwide. It has served a growing list of clients in Europe, Mexico, South Africa, the Middle East, Israel, India, Canada, and the United States with training, research, and educational projects and has provided specialized courses in space development for the International Astronautical Congress.

- GSI is an international education and research institute dedicated to the future development of space and especially new applications serving humankind.
- GSI was founded by space professionals with decades of international experience within national and regional space programs, the private sector, and world-class universities.
- GSI is designed to provide a wide range of services to the international space community whether in the form of on-site training, targeted research and development of space-related products or services, or consulting support.
- GSI is available to assist with specialized space training and especially "new space" entrepreneurial initiative training, education, and analysis needs.

The International Association for the Advancement of Space Safety (IAASS) was legally established on April 16, 2004, in the Netherlands, as a nonprofit organization dedicated to furthering international cooperation and scientific advancement in the field of space systems safety. In 2004 IAASS became a member of the International Astronautical Federation (IAF). In 2010 IAASS was granted Observer status at the United Nations COPUOS (Committee on the Peaceful Uses of Outer Space).

In accordance with the Association charter, the IAASS membership is open to anyone having a professional interest in space safety. Members can be physical persons, corporations, agencies, universities, institutions, and other professional associations.

The Association exists to help shape and advance an international culture of space safety (technical, organizational, and socio-political), which would contribute to make space missions, vehicles, stations, extraterrestrial habitats, equipment, and payloads safer for the general public, ground personnel, crews, and flight participants. The Association also pursues the safeguarding and sustainability of the on-orbit environment to allow unimpeded access to space by future generations as well as to address cosmic hazards of all types—including asteroids, comets, and solar storms.

The mission of the IAASS is to advance all forms of space safety study, research, and practical implementation. The association is committed, through the knowledge and dedication of its members, to advance space safety internationally. Goals include:

- Avoiding risk badly measured or willingly underestimated;
- Providing education and training in the field and providing necessary knowledge concerning space safety especially that not made available by others;
- Avoiding and preventing a lack of management commitment and attention to all aspects of space safety;
- Seeking to avoid a lack of personal accountability in the field of space safety, which can make people negligent;
- Advancing the science and application of space safety;
- Improving the communication, dissemination of knowledge, and cooperation between interested groups and individuals in this and related fields;
- Improving understanding and awareness of the Space Safety discipline;
- Promoting and improving the development of Space Safety professionals and standards;
- Advocating the establishment of safety laws, rules, and regulatory bodies at national and international levels for the civil use of space.

INTERNATIONAL®
SPACE UNIVERSITY

ISU

At its central campus in Strasbourg, France, and at various locations around the world, the ISU provides graduate-level training to the future leaders of the global space community. The university offers a 2-month Space Studies Program, a 5-week Southern Hemisphere Program, and a 1-year Masters program related to space science, space engineering, systems engineering, space policy and law, business and management, and space and society.

These programs give international graduate students and young space professionals the opportunity to learn while solving complex problems in an intercultural environment. Since its founding in 1987, the International Space University has graduated more than 3000 students from 100 countries (as noted in red in the map below), creating an international network of professionals and leaders. ISU faculty and lecturers from around the world have published hundreds of books and articles on space exploration, applications, science, and development.

Contents

About the Authors

Ram S. Jakhu is Director of the Institute of Air and Space Law, Faculty of Law, McGill University, Montreal, Canada, where he teaches and conducts research in international space law, law of space applications, law of space commercialization, government regulation of space activities, law of telecommunications and Canadian communications law, and public international law. He manages and directs a multi-million-dollar research and outreach program for space law and policy. He is a member of the Space Council of the World Economic Forum and a "Fellow" as well as the chairman of the Legal and Regulatory Committee of the International Association for the Advancement of Space Safety. In 2007, he received a "Distinguished Service Award" from the International Institute of Space Law for significant contributions to the development of space law. In 2016, the International Association for the Advancement of Space Safety awarded him the "Leonardo da Vinci Life-long Achievement Award." He is managing editor of the Space Regulations Library Series and member of the editorial boards of the *Annals of Air and Space Law* and of the *German Journal of Air & Space Law*. He served as a member of the board of directors of the International Institute of Space Law, 1999–2013, and as director, Centre for the Study of Regulated Industries, McGill University, 1999–2004, and as the First Director of the Master's Program of the International Space University, Strasbourg, France, 1995–1998. He is a widely published author and the editor of an award winning book on *National Regulation of Space Activities*. His academic degrees include B.A., LL.B. and LL.M. in International Law from Panjab University; LL.M. in Air & Space Law and Doctor of Civil Law in Law of Outer Space & Telecommunications (on Dean's Honor's List) from McGill University.

Joseph N. Pelton, Ph.D. is the former Chairman of the Board of Trustees and Vice President and Dean of the International Space University as well as the Director Emeritus of the Space and Advanced Communications Research Institute (SACRI) at the George Washington University. He is a member of the Executive Board of the International Association for the Advancement of Space Safety and former president of the International Space Safety Foundation. Dr. Pelton also served as director of the Accelerated Masters of Science Program in Telecommunications and Computers at the George Washington University from 1998 to 2005. Dr. Pelton was the founder of the Arthur C. Clarke Foundation and has served on its board of directors for several decades. He was also the founding president of the Society of Satellite Professional International (SSPI) and a member of the SSPI Hall of Fame.

Dr. Pelton is a widely published and award winning author with over 40 books written or coauthored or coedited with colleagues. His book *Global Talk* was nominated for a Pulitzer and won the Eugene Emme literature award. Dr. Pelton is a full member of the International Academy of Astronautics, an Associate Fellow of the American Institute of Aeronautics and Astronautics (AIAA), and a Fellow of the International Association for the Advancement of Space Safety (IAASS). He received his degrees as follows: B.S. from the University of Tulsa, M.S. from New York University, and a doctorate from Georgetown University.

Yaw O.M. Nyampong is working as Senior Legal Officer, Pan African University, African Union Commission, in Addis Ababa, Ethiopia. He also served as the Executive Director (Academic Associate) of the Centre for Research in Air and Space Law, McGill University, Montreal, Canada. From 2010 to 2013, he was a postdoctoral research fellow at the Faculty of Law, where his research focuses on the environmental aspects of space exploration and use, particularly how to address the problem of space debris.

Aside from his academic laurels and research experience, Dr. Nyampong has extensive practical experience in the field of air and space law, having worked as an international consultant on a number of air law-related assignments for the International Civil Aviation Organization and the World Bank Group. Dr. Nyampong has attended and spoken at numerous air and space law workshops and conferences around the world and has written and published several scholarly articles and book chapters on issues of contemporary relevance in the field.

He holds both a Doctor of Civil Law (DCL) degree and a Master of Laws (LL.M) degree in Air and Space Law from the Institute of Air and Space Law, McGill University, Montreal, Canada. He also holds a Qualifying Certificate in Professional Law from the Ghana School of Law (2000) and a Bachelor of Laws (LL.B) degree from the Faculty of Law, University of Ghana, Legon (1998). He is a member in good standing of the Ghana Bar Association and the Law Society of Upper Canada (Ontario).

1 Introduction

Scope

This book explores the exciting potential of mining space to obtain needed natural resources. It also examines the international and national legal and regulatory processes that would apply to this new and exotic endeavor. For almost 50 years—from virtually the start of the Space Age in October 1957—there have been practical applications of space technology. Today we have many programs that explore space via astronautic and robotic missions, and scientific probes to understand the chemistry, physics and workings of the universe, but from a practical perspective we also have application satellites. It is the application satellites that constitute our various tools and mechanical servants in the skies. In fact, the very name given by Galileo to the satellites he discovered circling Jupiter was *satelles*, the ancient Latin word for "servant." He chose this because he envisioned these satellites doing the bidding of the giant planet Jupiter.[1]

Today the practical uses of satellites have grown and grown. First there were telecommunications satellites (which now include broadcasting satellites, mobile communications satellites, search and rescue and data relay satellites, and so on). Shortly after the first communications satellites were deployed, there were also remote-sensing satellites, weather satellites, and navigation and timing satellites. We may soon have robotic repair and refueling satellites, solar power satellites, as well as increasingly sophisticated satellites for various types of defense and security operations.

However, something entirely new is on the horizon for space applications.

[1] Joseph N. Pelton and Scott Madry, Chapter 6, "Satellites Serving Humankind," in Joseph N. Pelton's and Angelina Bukley, *The Farthest Shore*, 2010, Apogee Books, Burlington, Canada.

© Springer International Publishing Switzerland 2017

R.S. Jakhu et al., *Space Mining and Its Regulation*,
Astronautical Engineering, DOI 10.1007/978-3-319-39246-2_1

This next major commercial space application may redefine the future of space activities to include major off-world activities. It would constitute a serious attempt to reclaim natural resources from space, and is called, quite simply, space mining. Some of those engaged in this activity also envision the processing of materials in space and even space manufacturing.

New Space Industries and Space Mining Ventures

The field of space applications and transportation has seen, particularly in the United States, the evolution of the term "new space." This term has been used in the space industry since the 1980s, when there were new entrepreneurial entries into the aerospace industry, which included Orbital Sciences (now known as Orbital ATK) and SpaceHab. Recent variations on the theme occurred just after the dot-com boom, when Internet technology and wealthy entrepreneurs such Paul Allen, Jeff Bezos, Robert Bigelow, Elon Musk and other "space billionaires" created new space companies to compete with well-established aerospace giants, and open new space markets (Fig. 1.1).

The advent of "new space" industries has already accelerated the development of space planes, space tourism industries, lower cost commercial vehicles for access to the International Space Station, and space servicing robotic systems. The question for the next decade is whether new space ventures will be able to successfully launch commercially viable space mining companies. The business model, of course, is just one of the issues. Technology, regulatory frameworks, and other factors could also dictate the outcome.

As a Tauri Group paper on the topic of new space industry has stated: "These companies believed that flat hierarchies, compensation schemes developed to enable bottom-up decision making, and evolutionary technology development models, based on experiences in software development, would transform

Fig. 1.1 Some of the space billionaires leading the current new space revolution. *From left to right*: Robert Bigelow, Jeff Bezos, and Paul Allen

space business."[2] The last 15 years in particular have seen the birth of a host of new commercial space industries. This has been significantly stimulated by the XPrize competition and a number of enterprising space entrepreneurs who see the challenges of creating entirely new space technology and businesses as exciting and personally rewarding. Space X, Scaled Composites, Virgin Galatic, Sierra Nevada, the XPrize and many other new space enterprises have truly changed the landscape of today's space industry. Often these new companies and their modern entrepreneurial approach to space, particularly as exhibited in the United States, have given new prominence to the term new space, which is usually "intended to differentiate developing businesses in the space industry from the traditional business structures and norms established during the space race." The Tauri Group analysis of this phenomenon has proposed that the typical characteristics of "new space companies" included the following:

- flatter and more flexible organizations
- entities that are consumer-focused, innovative, and willing to take risks
- organizations that are focused on new technology solutions.[3]

The significant rise of "new space" companies is particularly relevant to this book in that the few companies that have now been formed to pursue space mining and which are now planning to extract natural resources from outer space are clearly new space enterprises. These organizations, along the way, may contract with and involve conventional and well-established aerospace companies in their ventures—just as Deep Space Industries has contracted with Northrop Grumman for engineering and design studies. But clearly it is the small, flexible, risk-taking and unconventional companies that have to date taken the lead in space mining efforts. These organizations as described in the chapters that follow do indeed tend to have small and flexible organizations, the inclination to take risks and to focus on entirely new space technology, to have modest capitalization, and are impatient with restrictive regulations and legal constraints, resulting from a desire to move ahead quickly.

In the spacefaring nations outside the United States the situation is for the most part much different. In the rest of the world it is usually the governmentally backed and funded space agencies that are developing the technology and undertaking the research that might lead to future space exploration and natural resource extraction. Nevertheless there are indications that private initiatives concerning new space activities are beginning to spread around the world. In Chaps. 6–9 the status of relevant space activities related to possible future space mining activities are discussed.

What Natural Resources Are Found in Space and Where Are They?

There are many asteroids that have a high metallic content, and some of them contain precious metals and rare earth minerals. One of the new space mining

[2] Jason Hay*, Paul Guthrie†, Carie Mullins‡, Elaine Gresham§, and Carissa Christensen* Global Space Industry: Refining the Definition of New Space http://enu.kz/repository/2009/AIAA-2009-6400.pdf

[3] *Ibid.*

companies has estimated that the resource content of just a single asteroid to be nearly $200 billion. Other asteroids and the Moon have water, and the Moon contains not only water but a valuable isotope known as helium 3 that could be used as a fuel for a nuclear fusion reactor. One of the new space mining companies has made helium 3 mining on the Moon one of its prime objectives.

The problem with asteroid mining is that there are truly so many different possible candidates. And, in fact, the terminology can be a bit confusing. First of all near Earth objects (NEOs) are also sometimes referred to as near Earth asteroids, (i.e., they come within 0.03 astronomical units of Earth's orbit or within about 4.5 million km). These NEOs are also referred to as PHAs, or potentially hazardous asteroids. Comets are also potentially hazardous objects that can fly close to Earth. But their speeds within the inner Solar System are so fast that they are not candidates for space mining and are far more dangerous that asteroids because of their accelerated velocities.

Within this book there will be references to NEOs, NEAs, and PHAs, but they are essentially one and the same. Those interested in space mining are interested in tracking the orbits of these NEOs, NEAs, or PHAs as targets of opportunity. Space scientists, astronomers, and others concerned with cosmic hazards and planetary defense are interested in these asteroids to prevent a major disaster because these space rocks could create (and indeed have created) major damage by crashing into Earth.[4]

There is underway a concerted effort to document the orbits of NEOs that could both be a threat and a boon to Planet Earth. Involved in that effort are NASA, the Safe Guard Foundation in Italy, and the now the U. N. International Asteroid Warning Network (IAWN). The truth is that spotting potential candidates for space mining is not an easy task. There are literally millions of NEOs, and finding them, cataloging their orbits, assessing their mineral content, and determining if they could easily be approached and exploited for mining is a very large job indeed. NASA has been tasked by the U. S. Congress with locating all potentially hazardous asteroids that are 140 m in diameter or larger.[5] Despite the use of very capable in-orbit resources and drawing on a large number of ground observatories NASA has not been able to complete this mission in over a decade. There are tens of thousands of NEOs that are larger than 140 m in diameter, but perhaps a million that are 30 m in diameter and tens of millions that are 10 m or more in diameter. The NEOWISE (Near Earth Object Wide-range Surveyor Explorer) satellite, the planned Sentinel project by the B612 Foundation, the proposed NASA NEOCAM infrared telescope, plus computer-optimized ground observatories can contribute to the inventory of NEO asteroids. And many of these objects might potentially prove to be candidates for mining. It is likely, however, that only small satellite explorers can identify the size, characteristics, and chemical content of

[4] Joseph N. Pelton and Firooz Alidadi, editors, Chapter One: Introduction, *Handbook of Cosmic Hazards and Planetary Defense*, (2015) Springer Publishing, N.Y.

[5] Section 321 of the NASA Authorization Act of 2005 (Public Law No. 109-155), also known as the George E. Brown, Jr., Near-Earth Object Survey Act, neo.jpl.nasa.gov/neo/report2007.htm.

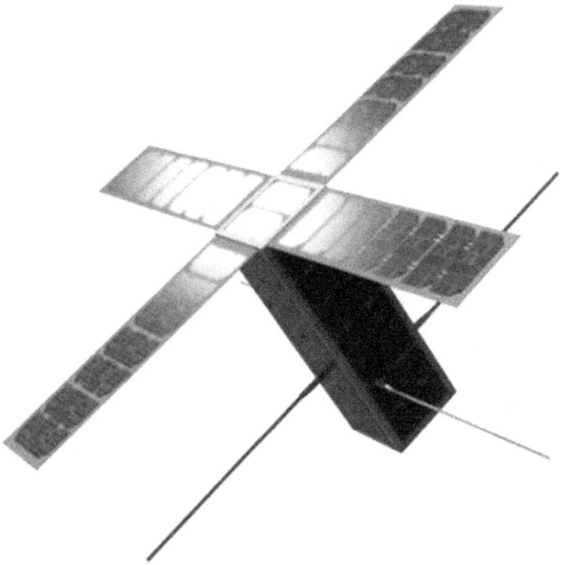

Fig. 1.2 Deep space industries concept of a nano-satellite to conduct exploratory missions (Graphic courtesy of deep space industries.)

potential candidates for space mining rather than the tools now in orbit or planned for deployment in the next few years. At this point both Planetary Resources Inc. and Deep Space Industries are planning to deploy such remote satellite explorers to find the prime candidates for space mining.

The Technology

This book contains two chapters that examine some of the key technologies needed for those seeking to engage in asteroid mining. These technologies involve more cost effective space transportation systems and precision navigation, remotely operated power systems and a variety of robotic capabilities that would be needed to carry out a variety of specialized operations associated with remote mining and natural resource recovery from outer space. In addition some of the new ventures are seeking to develop new, more cost effective ways to carry out remote surveillance with their nano-satellite explorers.

These and other unconventional approaches are briefly discussed in the chapter that reviews U. S. initiatives. Some of the proposed innovative technical approaches are currently being envisioned and in some cases even being developed. Figure 1.2 shows the concept for a "fire fly" nano satellite that could seek to obtain close up views of NEOs to determine if they are suitable candidates for natural resource extraction.

New Space and the Key Space Actors

Today there are only a few actors that are explicitly organized with the goal of engaging in space mining activities. These are Planetary Resources Inc., Deep Space Industries, and Shackleton

Energy. But beyond these commercial organizations there are many others involved in commercial transportation systems, precision space navigation, space robotics operations, and space-based power systems that are developing technology and systems that are relevant to the ultimate success of space mining ventures. In addition there are many space agencies as well as private foundations (i.e., B612 Foundation and Safeguard Foundation) land-based observatories and universities and research centers that are carrying out important work that are of considerable value to those planning to engage in space mining activities. Some of the more important agencies, research centers, observatories and universities conducting work in this field are listed in the appendices.

The Legal and Regulatory Context: Today and Tomorrow

There are a number of international treaties, conventions and international regulations that are relevant to efforts to undertake a space mining operation in the future. Those that are most germane include the following, as identified by their full titles below.[6]

- Treaty on Principles Governing the Activities of States in the Exploration and Use of Outer Space, Including the Moon and Other Celestial Bodies (October 10, 1967; this treaty is generally referred to as the "Outer Space Treaty," or OST);
- Convention on International Liability for Damage Caused by Space Objects (October 9, 1973; this treaty is generally referred to as the "Liability Convention");
- Convention on Registration of Objects Launched into Outer Space (September 15, 1976; this treaty is generally referred to as the "Registration Convention");
- Convention on Early Notification of a Nuclear Accident (October 27, 1986; this treaty will be referred to as the "Notification Convention");
- Convention on Assistance in the Case of a Nuclear Accident or Radiological Emergency (February 26, 1987; this convention will be referred to as the "Assistance Convention").
- Principles on the Use of Nuclear Power Sources in Space.

These and other efforts at setting up international law as well as so-called "soft law" and areas of "transparency and confidence building measures" will be discussed in the later chapters. A part of this discussion is whether special international rules and treaty considerations apply to the Moon and other "celestial bodies" and whether these substantial cosmic bodies are different from asteroids and space rocks.

Beyond this fundamental issue there are nevertheless a number of procedures and guidelines regarding the registration and licensing of launched objects, the use of nuclear power sources and isotopes in space, liability consideration, and considerations that any government or commercial entity would need to

[6] Gary L. Bennett, "Proposed Principles on the Use of Nuclear Power Sources in Space," http://fas.org/nuke/space/propprin.pdf.

comply with if it were to proceed with space mining operations.

The discussion in this book of relevant outer space treaties, conventions and guidelines is certainly not intended to be definitive, but this analysis does seek to identify the most important and salient international legal and policy issues. In short book this tries to set forth the main issues and indicates the international provisions that are now in place and pertinent pending proposals either within the U. N. Committee on the Peaceful Uses of Outer Space (COPUOS) or other relevant forums that have begun to consider these issues.

The Longer Term Perspective

The tendency in almost any institution that creates new laws, conventions and especially international treaties is: "Let's wait until there is a clear problem to be addressed and then we will address it." The major issues associated with atmospheric and oceanic pollution, climate change, orbital space debris, destruction of the rain forests, etc., are some cases in point. These are now major contentious issues where an effective international response to these problems has proved to be very difficult and expensive to address. Such problems can become severe over time. Ameliorative action can result in major restrictions on industries and even governmental programs. Severe economic penalties, fines or limits on usage can be imposed to limit such programs. If these issues had been addressed decades ago, the economic and political costs could have been substantially minimized. Although opportunities have been lost with issues such as orbital debris in Earth orbit, it might be possible to create reasonable and proactive rules and regulations with regard to space mining so that the economic and even social benefits of space mining can be realized, with future dangers and negative aspects avoided.

Arthur C. Clarke in his book *Rendezvous with Rama* anticipated the possible future dangers of a catastrophic impact on Earth by a rogue asteroid. In this novel, set in 2077, he discussed the formation of a global "Project Safeguard" against an asteroid strike.[7] However, the United Nations in December 2013 agreed to the formation of an International Asteroid Warning Network (IAWN) as well as a Space Mission Planning Advisory Group (SMPAG).[8] The United Nations, under intense lobbying by astronauts, astrophysicist and members of Team 14 from COPUOS was able to take on this issue 64 years ahead of Clarke's schedule in his science fiction novel. In an area as forward looking as space exploration and applications, it would be a hopeful sign if we were to be able to anticipate problems that might emerge from space mining activities and seek to address those problems proactively, rather than in the usual fashion—i.e., after the fact.

It would seem prudent for the U. N. Committee on the Peaceful Uses of Outer Space and its Working Group on the Long Term Sustainability of Space Activities to seek to develop practical

---

[7] *Rendezvous with Rama,* http://www.goodreads.com/book/show/112,537.Rendezvous_with_Rama.

[8] Leonard David, "Dealing with Asteroid Threats: UN Completes First Planning Phase" Space.com http://www.space.com/28,755-dangerous-asteroids-united-nations-team.html (Accessed: August 22, 2015).

safety guidelines to avoid dangers to Earth and to allow space mining to proceed while minimizing future problems of an environmental or other nature. Organizations such as the Secure World Foundation (SWF), the International Association for the Advancement of Space Safety (IAASS), and structured meetings among space agencies might be solicited to help provide advice and counsel on useful guidelines and procedures that everyone engaged in this type of activity might usefully follow.

Although it may seem quite unconventional, and even quite exotic, this book has indeed looked ahead through the eyes of science fiction writers to envision a future where there is actually space mining and astro-engineering and manufacturing at space colonies in order to see what future problems—and solutions—can be anticipated so that action can be taken now to anticipate and avoid future problems.

Structure and Purpose of This Book

This book, although brief in its presentation, seeks to be comprehensive in scope. Thus it examines the types of technologies that space mining ventures will need to develop and reliably operate if they are to succeed in actually creating a "new space" venture that is viable. It analyzes the various relevant governmental and commercial programs that are now being carried out that are antecedent to future space mining operations. It examines programs in the United States, in Europe, Canada and Australia, in Asia (China, Japan and India), in Russia and in the rest of the world. Although there is currently the most activity in the United States, there are significant efforts to create new space transportation systems and key new space technologies being developed in other parts of the world. If U. S.-based activities move ahead, clearly other initiatives will closely follow from around the world. The analysis of space-based activities around the world will seek to consider not only technical, operational, business, economic, and financial capabilities and opportunities but also the relevant legal and policy framework issues as well.

This book will examine the existing international treaties, conventions, and other "soft law" instruments or "codes of conduct" that might apply to efforts to engage in space mining in future days. The issue of space mining is complicated by the fact that this is an important issue of "degree." If one were to go to the Moon and set up mining operations, the activity would seem to be clearly covered by the Outer Space Treaty and the Moon Agreement. On the other hand many tons of "space dust" falls to Earth every day. An organized effort to reclaim this resource would not be thought to come under the heading of space mining. In short one of the difficulties seems to be considering what size of NEO might legitimately be considered a "celestial object" and perhaps subject to some sort of international regulatory process and what might be considered so small that it is "space junk."

Also there are the other issues of cosmic hazards and planetary defense. On one hand an NEO might be considered a candidate for reclaiming for space mining, but on the other it might be considered a potential danger to the planet.

Thus those that propose to engage in space mining could either endanger the world or alternatively rescue it from a deadly impact. This is a separate policy and regulatory issue from the one of who "owns" NEOs and becomes one of who might be permitted to engage in "protective" actions.

Thus the goal of this book is to examine the technologies, the various incipient efforts to engage in space resources identification and natural resource extraction around the world, and then to explore the legal and regulatory issues surrounding such efforts.

To assist this effort there is a glossary of terms provided at the end of the book that explains acronyms and key terms and phrases. Also key legal, regulatory, and official policy materials that are relevant to activities involving the harvesting of natural resources from outer space are provided as useful references in the appendices as well.

2 The Importance of Natural Resources from Space and Key Challenges

Earth is a 6 sextillion-ton globe that contains a wealth of resources. These natural resources can be extracted from the atmosphere, the oceans and the ground. If these resources are used wisely and in a sustainable fashion they should be able to be recycled and used over and over again. However, as the global population has increased from 800 million in 1800 to 1.8 billion in 1900 to 6.3 billion in 2000 and to well over 7 billion today, the demand for fossil fuels for transportation and energy needs has expanded greatly, as well as the need for various metals and other rare earth materials. Modern civilization with its complex infrastructure, burgeoning population, and surging urban complexes that will within three decades contain 70 % of all humanity, is in need of significant re-engineering to adjust to twenty-first century realities. In a world with perhaps a 100 megacities of more than 10 million people each, significant elements of climate change, major environmental shifts, and natural resource needs, the world as we know it today will have to significantly change, or it will no longer be sustainable.[1]

Some of the most significant changes that will have occurred by the time of the twenty-second century are the following:

- Human population stabilization. The exponential growth of human population that has characterized the eighteenth, nineteenth, and twentieth centuries will likely give way to a zero population growth profile in the twenty-first century at a level between 10 and 12 billion people. Even at this level demands on natural resources, climate change issues, and food and energy needs will be challenging. Continued growth within the energy, food and natural resource limits is not sustainable. This new world that is 70 % or more urban will

[1] Joseph N. Pelton and Peter Marshall, *Megacrunch: Ten Survival Strategies for 21st Century Challenges* (2010) PMA Associates, London, United Kingdom.

© Springer International Publishing Switzerland 2017

R.S. Jakhu et al., *Space Mining and Its Regulation*,
Astronautical Engineering, DOI 10.1007/978-3-319-39246-2_2

be far more vulnerable to the loss of critical infrastructure.[2]

- Shift to sustainable sources of energy. Twenty-first century energy systems will be weaned off of fossil fuels, and there will be a gradual shift to solar, wind, geothermal, hydroelectric, tidal, and nuclear fusion-based energy systems. In making this transition, efforts to decentralize power supplies can add to sustainability and survivability of urban communities.
- Climate change stabilization. The efforts to stabilize Earth's climate and slow down natural and human-fueled climatic shifts will increase. It may be recognized in coming decades that space-based solutions—such as a space-based shield at Lagrange Point One or space-based heat pipes or other mega-engineering projects may be needed to provide ultimate answers. This is not only an issue of human avoidance of mass extinction but also an issue of the preservation of a wide range of animal and plant life.
- New patterns and forms of urbanization. The last two centuries has brought about a vast shift in global patterns of urbanization. The shift from less than 5 % urban to over 70 % urban will have occurred in less than 200 years. This change has been driven first and foremost by jobs and employment. In an age of optical and electronic computer and telecommunications networks, increasing super automation and robotics, and a service economy, the current patterns of super-urbanization are no longer necessary and indeed are dangerous in terms of over concentration of populations. Increasingly, there will be shifts in patterns of urbanizations. New "meta cities" that are geared to twenty-first century telecommunications, networking, energy and transportation systems will serve to relieve pressures on megacities of 10 million or more people. Telework can help ease the problems of super urbanization and over concentration of population.[3]
- Space systems for planetary defense. Over the next few decades planetary defenses will shift from threat detection to threat protection. Space systems can be developed to alleviate threats from solar storms, a weakening geomagnetosphere, potentially hazardous asteroids and comets, and even threats from runaway orbital debris. Ultimately space programs and systems will be recognized not as luxuries but rather as essential capabilities to preserve the human race against mass extinction. Capabilities developed to support space mining will contribute vital technologies to achieve effective planetary defense as well as provide vital resources for the future.
- Exhaustion of planetary natural resources and a new extra-terrestrial based economy. Secretary of State John Hay once famously said: "The Mediterranean is the ocean of the past, the Atlantic is the ocean of the present, and the Pacific is the ocean of the

[2] Indu Singh and Joseph N. Pelton, *The Safe City: Living Free in a Dangerous World* (2013), The Emerald Planet, Washington, D. C., pp. 193-198.

[3] *Ibid.*, pp. 215-233.

future."[4] And over time the global economy has expanded to make this prediction a reality. Soon the economies of China, India, Indonesia, and Japan plus the smaller countries of Singapore, Taiwan, Republic of Korea, Thailand, etc., will outstrip those of the United States and Europe. As these developing economies grow more prosperous and demand for natural resources continue to grow the availability of natural resources will become an increasing problem. In looking to the future it might be well to consider the predictions of Ray Kurzweil and his forecast of the coming "Singularity."[5] Or perhaps to consider the vision of Peter Diamandis in forecasting a future that is increasingly based on an extra-terrestrial and space-based economy. The value of platinum-rich asteroids has been estimated at levels ranging from $200 billion up to even a trillion dollars. Of course the future of our energy needs rest with the Sun. With ever growing space-based capabilities and the cost of going into space ever decreasing, the reality of a space-based economy becomes more realistic every year.[6]

Gauging the Future

The future is most often viewed through a rearview window. And for millions of years of biological evolution the past was indeed often prologue to the future. With the advent of technology, computer and communications networks, artificial intelligence, robotics, and the ability to go into space, the rate of change in human civilization has increased exponentially. In Fig. 2.1 the "Super Month" graphic compresses the time since the age of the Southern Ape Man into a 30-day period where every second represents 2 years. In Super Month time the age of farming and permanent settlements represents the last hour and a half of the last day of the month, the Renaissance is the last 4 min, and the Industrial Age is 2 min until midnight. The age of computers, cell phones, television, bio-engineering, megacities, space launches and spandex—all the elements of modern life we take for granted—represent only the last 20 s of "Super Month" time.

This graphic serves to demonstrate that judging future societal needs on past experience is a seriously flawed concept. The future needs of human civilization in terms of energy, housing, transportation, water, natural resources, jobs and employment, and security are dramatically different that they were ever before. As an illustrative example, it has been artfully noted that it would be far easier for Moses to come and live in the times of Napoleon and Thomas Jefferson than for someone living in the eighteenth century to come forward and live in today's world of advanced technology.

When Peter Diamandis talks of "abundance" and Ray Kurzweil talks of

[4] John Hay quote on the Pacific as the ocean of the future https://books.google.com/books?id=5P9bgGxfYKUC&pg=PA118&lpg=PA118&dq=John+Hay+quote+on+Pacific+ocean&source=bl&ots=8Tb4vBEDMm&sig=k6wOGDKzmnb3DHqVonBrhSE9AVA&hl=en&sa=X&ved=0CDMQ6AEwA2oVChMI35rYq8q_xwIVU4MNCh1YgAjL#v=onepage&q=John%20Hay%20quote%20on%20Pacific%20ocean&f=false

[5] Ray Kurzweil, How to Create a Mind (2012), Penguin Group, London, U.K.

[6] Peter Diamandis, *Abundance: The Future is Better than You Think*, (2012) Free Press, New York.

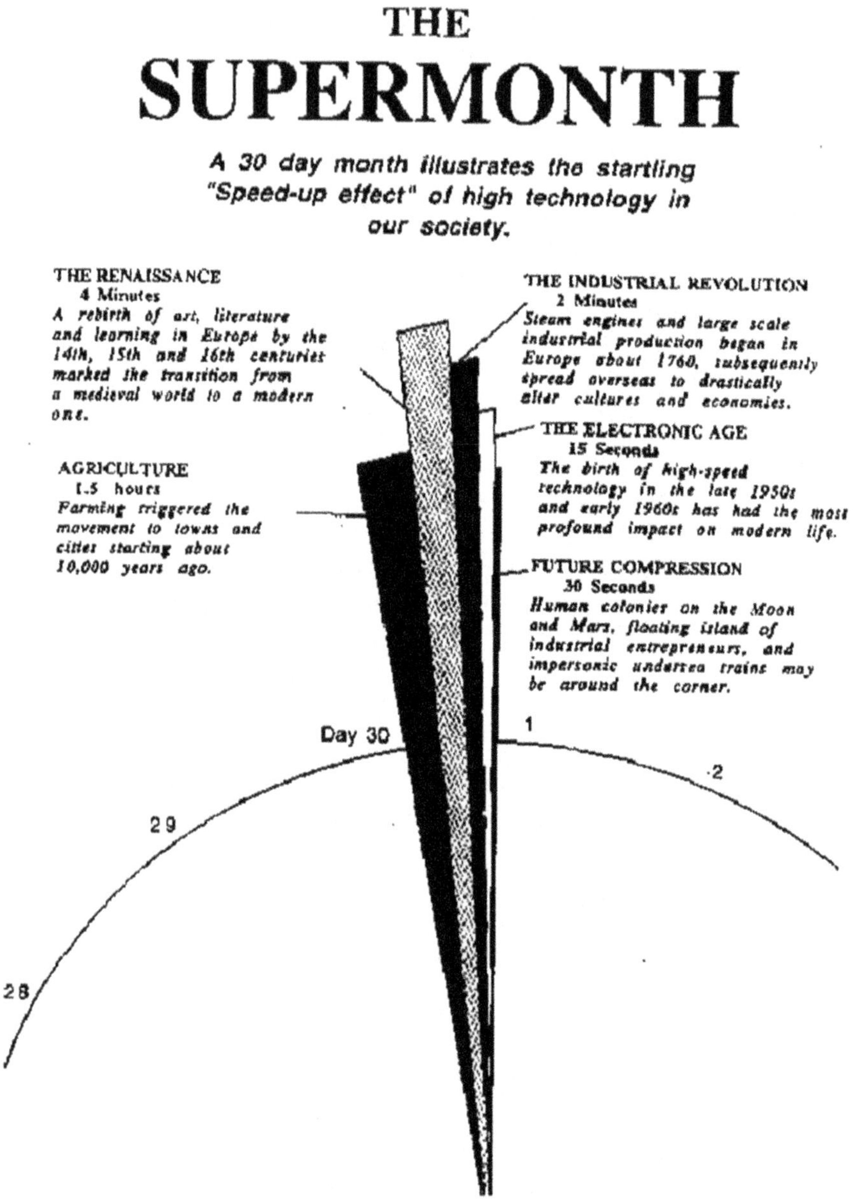

Fig. 2.1 This figure shows future compression and accelerating innovation (Illustration courtesy of the author.)

"the singularity," they are suggesting a world that is far different than we ever experienced before. They envision a world in which we cohabitate with robots that are as smart as humans and possess sophisticated "thinking" skills, space probes that can be used to bring us new resources and clean energy from

Fig. 2.2 The volume of Earth's accessible water in comparison to the volume of the world (Graphic courtesy of the Sierra Club.)

space and also protect our critical infrastructure from cosmic hazards.

Coping with the Scale and Complexity Problem

The land area of the entire world is 148.94 million sq. km (or 57.506 million sq. miles), and its water area is 361.132 million sq. km (or 139.434 million sq. miles). About half of that land area is truly viable for year- round habitation when one eliminates most parts of Antarctica, the Arctic north, Siberia, the most dangerous mountain ranges and the most arid desert regions. Rising sea levels will further decrease available land areas. When one divides about 75 million sq. km by 10 billion people (or about 133 people people/sq. km) it becomes clear that rising global population and shrinking land areas and exhaustion of many types of natural resources—especially potable water—will be a growing problem.[7] Figure 2.2 shows the volume of water in the world in comparison to the total volume of Earth. This graphic helps us to realize just how small the amount of potable water that is truly accessible today in comparison to a rising global population actually is.

Figure 2.2 underscores the issue of just how difficult it will be to continue to provide key resources especially to major urban centers as global population continues to grow. And this is not

[7] A profile of the world https://www.google.com/search?sourceid=navclient&ie=UTF-8&rlz=1T4VSND_enUS583US595&q=What+is+the+Land+area+for+the+world%3f (accessed August 24, 2015).

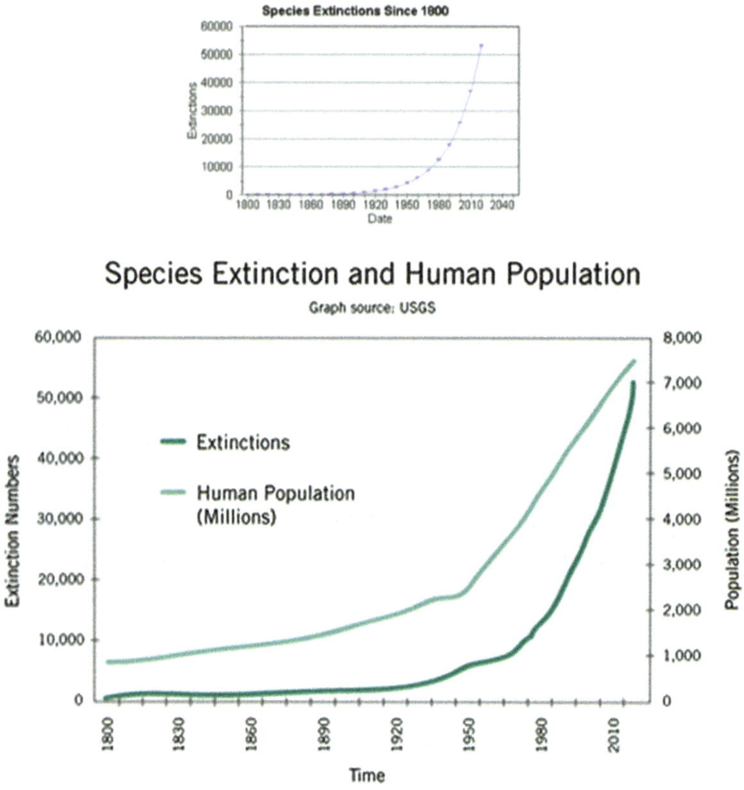

Fig. 2.3 Rise of extinction levels of species vs. human population growth (Graphic courtesy of the U. S. Geological Survey.)

just a question of sustaining human needs for water and natural resources. It is also a matter of sustaining endangered species of flora and fauna. The United Nations had done an analysis that shows the loss of species since 1800 and projections for the future show a very disturbing trend.[8]

The graphs in Fig. 2.3 that come from the U. S. Geological Survey seem to show a relationship between the rapid growth of the global human population in recent times and the increasing rate of extinction on species.

The future availability of petroleum products and water is most often mentioned in studies of future resource scarcity, but broader studies have shown that the world by the mid twenty-first century will have many shortages. The following results from a detailed Global Nonrenewal Natural Resources (NNR) study came up with the following

[8] Gail Tverberg, "A Look at the Latest UN Predictions on Natural Resource Consumption, http://oilprice.com/Energy/Energy-General/A-Look-At-The-Latest-United-Nations-Predictions-On-Natural-Resource-Consumption.html.

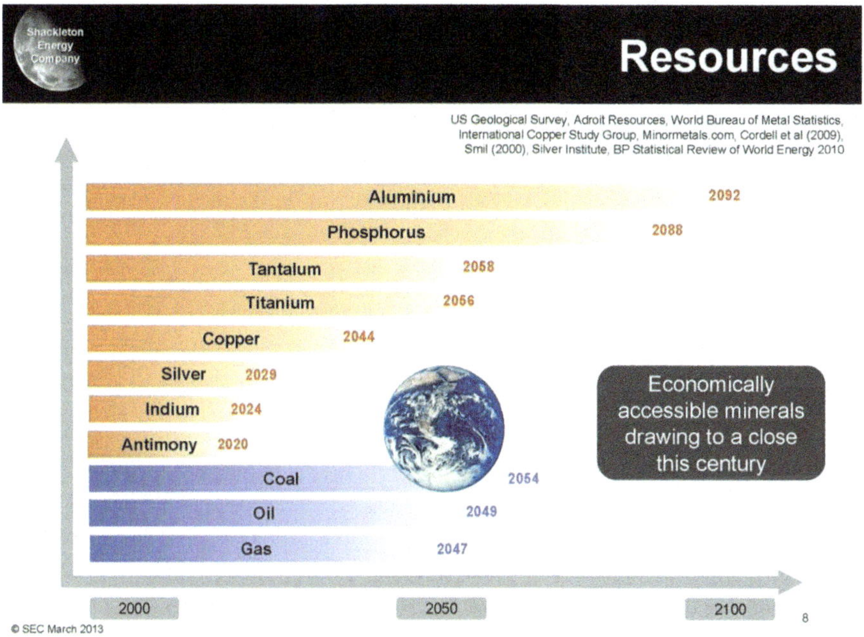

Fig. 2.4 Projected shortages of economically accessible minerals (Graphic courtesy of Shackleton Energy Company.)

results, as shown in Fig. 2.1.[9] Although these results might vary somewhat from year to year based on economic downturns or upturns, the overall trend toward increasing shortages is clear. The upward mobility of the populations in China, India, Indonesia, and other newly industrialized companies suggest that up to three times more consumer demand for products and energy will be present by the middle of the twenty-first century. Only recycling and new energy sources can meet the great bulk of this burgeoning demand. Meeting the demand for natural resources has been identified as a problem by many that have researched this problem. The projections of shortages in the future are presented in Fig. 2.4 and in even greater detail in Fig. 2.5 are certainly of concern. As Chris Clugston's detailed analysis of this subject has concluded: "Global Non Renewable Natural Resource (NNR) scarcity will intensify going forward, as global economic activity levels, economic growth rates, and corresponding NNR demand return to their pre-recession levels; and global NNR supply levels continue to approach and reach their geological limits."

Yet the prospect of space mining can provide new options. A modest near-Earth asteroid rich in platinum, approximately spherical in shape and 30 m in diameter would constitute a volume of 4500 cu. m and represent a mass of

[9] Chris Clugston, "Increasing Global Nonrenewable Natural Resource Scarcity—An Analysis http://www.resilience.org/stories/2010-04-06/increasing-global-nonrenewable-natural-resource-scarcity%E2%80%94-analysis.

Pre-recession (2000-2008) Global NNR Scarcity Summary

Extremely Scarce (6)	Very Scarce (21)	Moderately Scarce (22)	Marginally Scarce (1)	Not Scarce (7)
Bromine	Aluminum	Antimony	PGM	Arsenic
Gold	Bauxite	Beryllium		Barite
Mercury	Cadmium	Bismuth		Boron
Tantalum	Cement	Coal		Diamond
Tellurium	Chromium	Cobalt		Garnet
Thallium	Copper	Gallium		Lithium
	Fluorspar	Germanium		Niobium
	Magnesium Compounds	Graphite		
	Molybdenum	Gypsum		
	Natural Gas	Indium		
	Nickel	Iron Ore		
	Nitrogen (Ammonia)	Lead		
	Oil	Lime		
	Phosphate Rock	Manganese		
	Potash	Salt		
	REM	Silicon		
	Rhenium	Silver		
	Selenium	Soda Ash		
	Strontium	Tin		
	Sulfur	Vanadium		
	Tungsten	Zinc		
		Zirconium		

Fig. 2.5 An analysis of non-renewable natural resources reaching their geographic limits (Note: This chart is derived from information included in an article by Chris Clugston entitled "Increasing Global Nonrenewable Natural Resources—An Analysis," *The Oil Drum,* April 6, 2010.)

perhaps 5000 metric tons. If one assumed that this asteroid was 50 % platinum, then its value at current world market prices would be on the order of $90 billion. Even if the asteroid recovery mission and refinement costs ran to $5 billion and even if some of the proceeds were to go into some sort of global commons development or ecological fund, just a single such mission would produce many billions of dollars in profits. This may represent an extreme example, but there are over a million PHAs that are on the order of 30 m. The key in the early days of space mining would be to identify high-value targets.

A 50-m PHA would be over 4.6 times more massive in volume and content and would be incredibly valuable if it contained precious metals or rare earth materials such as iridium, rhodium, ruthenium, palladium, or osmium. In contrast, the economics would be much more difficult in the case of PHAs with less valuable natural resource contents. An asteroid with 70 % nickel and molybdenum content and 50 m in diameter would have something like a market value of only about $200 million based on current market prices of $13,000 a metric ton for molybdenum and $10,000 a metric ton for nickel. This much lower valuation would call for space mining transport equipment of the longer term future that could be used over and over again. It would also likely mean systems that ran off of solar and electric propulsion systems.

It is important to note that space mining activities can be cost effective for recovering at least rare metals, but it is also noteworthy that even hydrogen, oxygen or water or other volatiles in space can also be valuable. The

Planetary Resources website states: "In orbit, spacecraft propellant is a multi-billion dollar industry with each pound of fuel worth more than an equivalent pound of gold on Earth. Certain asteroids are loaded with hydrogen and oxygen, the components of rocket fuel. These asteroids can provide a fuel source that is 100 times closer energetically to Earth orbit, and thus far less expensive, than the Apollo-Era "bring-everything-with-you" propellant used today."[10]

But the space mining industry can also aid in producing and perfecting new technologies that could assist with other types of space missions, or produce innovations that can find useful implementation right here on Earth. Space mining activities will be seeking to develop new and more cost effective robotics missions, advanced navigation and precision maneuvering in space, improved space situational systems, lower cost satellite manufacturing techniques, and improved power systems, including higher efficiency photovoltaic cells and quantum dot technology.

Of course the most important contribution could well be more cost effective space transportation systems such as solar-powered electric propulsion systems. If one could develop transport systems that are largely multi-use that can be used over and over again, they could also be employed to boost cost effective solar power satellites into orbit.

Likewise if space mining enterprises can develop low cost satellites that could produce at lower cost and in high volume via 3-D printing, such as Planetary Resources is now developing, this could be quite significant. Such techniques could also find application in communications, precision satellite navigation, and remote sensing constellations and on other space missions. Clearly low cost remote surveying and reconnaissance satellites are currently the top priority for space mining ventures, and Fig. 2.6 shows the prototype small satellite that Planetary Resources Inc. together with 3D Systems is currently developing. This Arkyd-300[3] satellite bus configuration as pictured below with its efficient torus shape holds the propellant and provides the structure for the satellite. The fact that the satellite can be "manufactured" via 3D printing, of course, greatly reduces its production cost. One of the characteristics of the new space mining companies is that they have typically recruited partners that can help them develop these new types of technology. They have also been skillful in winning contracts from NASA for research and development work.[11]

Coping with Legal, Regulatory and Standards Problems

The current state of the space mining incipient industry is that they have been far more adept at identifying the scientific, engineering and technological challenges to be faced and pursuing systems solutions than they have been at

[10] Planetary Resources overview, http://www.planetaryresources.com/company/overview/#why-asteroids (accessed August 24, 2015).

[11] 3D Systems and Planetary Resources Announce Investment and Collaboration, June 26, 2013. http://www.planetaryresources.com/2013/06/3d-systems-and-planetary-resources-announce-investment-and-collaboration/.

Fig. 2.6 (*Left to right*) Peter Diamandis, Chris Lewicki, and Steve Jurvetson of Planetary Resources unveiling the Planetary Resources 3D-printed satellite in February 2014 (Image courtesy of Planetary Resources.)

addressing what might be called the legal, regulatory, and standards problems that this new type of enterprise entails.

The only "established" international law that has widespread acceptance is the Outer Space Treaty.[12] The so-called Moon Treaty has only a few signatories and has not been signed by many space powers. Other provisions such as the Liability Convention is also relevant as well as various efforts to define provisions regarding the use of nuclear systems in space. The most relevant parts of the Outer Space Treaty are Articles 1 and 2 that state:

> ***Article I***
> *The exploration and use of outer space, including the moon and other celestial bodies, shall be carried out for the benefit and in the interests of all countries, irrespective of their degree of economic or scientific development, and shall be the province of all mankind.*
>
> *Outer space, including the moon and other celestial bodies, shall be free for exploration and use by all States without*

[12] Treaty on Principles Governing the Activities of States in the Exploration and Use of Outer Space, including the Moon and Other Celestial Bodies http://www.unoosa.org/oosa/en/ourwork/spacelaw/treaties/outerspacetreaty.html.

discrimination of any kind, on a basis of equality and in accordance with international law, and there shall be free access to all areas of celestial bodies.

There shall be freedom of scientific investigation in outer space, including the moon and other celestial bodies, and States shall facilitate and encourage international co-operation in such investigation.

Article II
Outer space, including the moon and other celestial bodies, is not subject to national appropriation by claim of sovereignty, by means of use or occupation, or by any other means.

Despite these provisions there are at least three private entities seeking to engage in space mining activities. One of the key questions discussed in the book is exactly what the definition of a celestial body is and whether a small asteroid of which there are millions constitutes a celestial body. Clearly such asteroid mining of small and potentially hazardous bodies would seem much more permissible under the Outer Space Treaty than efforts to establish national or private colonies on the Moon or to undertake mining operations on the Moon.

There are currently a number of efforts underway to seek to clarify the future prospects for space mining activities—whether they might be governmental or private ventures—and to address the future legal and regulatory status of such undertakings. This effort includes the McGill University Air and Space Law Institute "Global Space Governance Study" that devotes a chapter to this topic. From a practical rather than a legal viewpoint it would seem that the projects by Planetary Resources and Deep Space Industries seem to have less of a legal, regulatory or standards challenge to their proposed efforts than Shackleton Energy, which is focused on mining on the Moon.

Conclusions

The website of Planetary Resources contains the grand statement that inhabitants of Earth are currently limited to the finite resources found on our planet, but that we do not have to be limited to this fate over the longer term future. There indeed may be a need to create structures in space to defend our planet against extreme solar storms and new types of space infrastructure to beam new forms of clean energy down to the ground. The future is clearly not what it used to be. New space industries could indeed change our future—perhaps for the better or perhaps for worse. The "future compression" innovations that are bringing future realities to the fore at an ever more rapid pace in "Super Month Time" are ever more apparent each day. These changes will require institutional and legal responses in a more proactive manner. The ever growing innovations that include a wide range of new space initiatives will continue to bring the future into our lives with an urgent necessity.

3 Transportation Systems and Targeting Locations for Space Mining

Just as there a number of different approaches to the robotics systems for space mining there are different approaches to space transportation systems for space mining.

The first key to developing space transportation systems for space mining involves determining the intended locations for such operations. At this time there are, in a generic sense, two prime candidates for nearer term space mining. First and most obviously are the Moon and Earth's "other Moon," namely the 5-km diameter asteroid 3753 Cruithne.[1] Second, there of course selected near-Earth asteroids that are located in suitable orbits and that also have mineral and metal contents that are attractive for mining. Clearly the Moon is a celestial body, and space dust, of which many tons fall to Earth, is not. Exactly what constitutes a celestial body is open to discussion.

The various types of NEAs are divided according to their various types of orbits. Some travel near Earth entirely inside its orbit. Some travel near Earth entirely outside its orbit, and others (the most dangerous) intersect with Earth's orbit twice a year. Some of these are quite elliptical, but the ones that are most circular are in many cases prime candidates for space mining since these are often easiest to reach and easiest from which to bring mined materials back to Earth. The following, chart prepared by Jet Propulsion Lab scientists, explains the taxonomy of the major types of NEA orbits. Fortunately the Apollo-type NEAs that are often prime candidates for space mining are also the most numerous. Unfortunately these type NEAs can also be identified as candidates to be Earth impactors. One might argue in this sense that mining Apollo NEAs could help to eliminate the threat of asteroids that could one day cause a major catastrophe on Earth (Fig. 3.1).[2]

[1] "Duncan Forgan, The 'Second Moon' You Didn't Know Earth Had," Discover Magazine, March 2, 2015. http://blogs.discovermagazine.com/crux/2015/03/02/earth-second-moon/#.VV9I8U9VhHw.

[2] "What are Atira, Atens, Apollos, Amors," JPL Frequently Asked Questions eo.jpl.nasa.gov/faq/#aten.

© Springer International Publishing Switzerland 2017
R.S. Jakhu et al., *Space Mining and Its Regulation*,
Astronautical Engineering, DOI 10.1007/978-3-319-39246-2_3

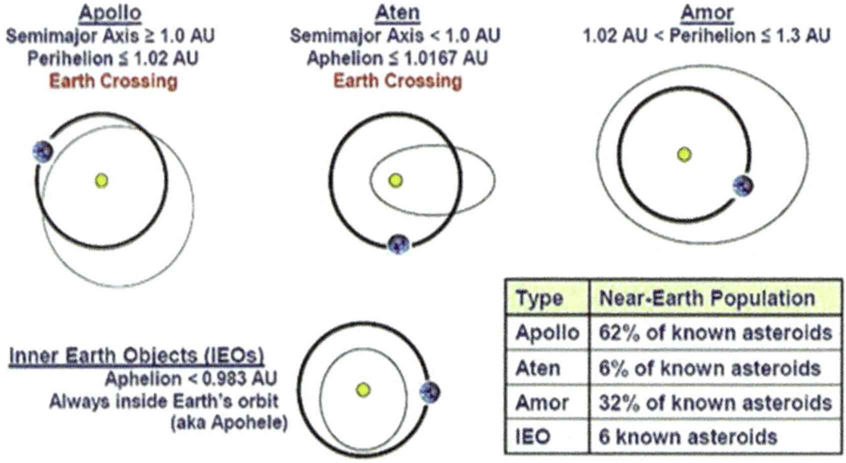

Type	Near-Earth Population
Apollo	62% of known asteroids
Aten	6% of known asteroids
Amor	32% of known asteroids
IEO	6 known asteroids

Fig. 3.1 Various types of near Earth asteroids in relation to Earth orbit (Graphic courtesy of NASA-JPL.)

The AUs in the chart above stand for astronomical units, which is the distance between the Sun and Earth and is precisely defined as 149,597,870,700 m or some 93,000,000 miles. The reference to 0.02 AU, for instance, would signify a distance very close to 3 million km.

The current listing of possible candidates for NEAs that might be considered for space mining has been compiled by Planetary Resources and is shown on its website. These NEAs are at this stage only provided as possible examples and are based primarily upon the modest increase in velocity (i.e., 4–5.5 km/s) that would be needed to reach such candidate asteroids from low Earth orbit. What is perhaps most interesting about the NEAs listed is that some of these have been discovered as recently as in 2014 and 2015 (Table 3.1).[3]

Beyond this modest list of candidate NEAs for space mining, however, there is the Jet Propulsion Laboratory (JPL) small-body database browser that provides a much larger inventory of small bodies that have been detected and might become future candidates for asteroid mining.[4]

Planetary Resources and Zooniverse, which is a citizen-based space project founded in June 2014, has devised a web-based activity known as Asteroid Zoo (www.asteroidzoo.org).[5]

This activity is keyed toward allowing students, citizen scientists and space enthusiasts to search for previously undiscovered asteroids and report them on the web for further investigation. The project uses observations from the Catalina ground observatory to seek out the orbits of NEAs. This innovative program championed by Planetary

[3] Asteroid targets of interest: http://www.planetaryresources.com/asteriods/#asteroids-targets.

[4] JPL Small Body Database Browser: http://www.planetaryresources.com/asteriods/#asteroids-targets.

[5] Asteroid Zoom: Hunt for Resource-Rich Asteroids http://www.asteroidzoo.org/.

Table 3.1 Candidate asteroids for space mining as identified by planetary resources

Possible candidate asteroids for space mining					
Name	Diameter	Delta velocity from LEO to reach	Time to reach/ mission cadence	Type of orbit	Arkyd space mission to examine
2014EJ24	About 100 m? (60–135 miles)	4–4.5 km/s	9 months/1 year	Apollo	Yes-Sept 2016
2014 SC324	About 60 m? (40–90 miles)	Too small and too elliptical orbit to be of interest. Narrowly missed Earth in 2014		Aten or elliptical	
1999JU3	840–900 m	4.5–5.0 km/s	11 months/4 years	Apollo	Japanese Hayabusa-2 explored-carboneous
2002TC70	About 300 m? (188–420 miles)	4.5–5.0 km/s	6 months/3 years	Apollo	
2011 CG2	About 225 m? (136–305 miles)	5–5.5 km/s	10 months/4.5 years	Apollo	
2001 QC134	About 300 m (270–330 miles)	5–5.5 km/s	6.0 months/5.9 years	Apollo	Stony
2013 PA7	85–190 m	4.5–5.0 km/s	1.8 months/5 years	Amor	
2008 HU4	About 10 m (6–13 m)	4.5–4.0 km/s	2 months	Apollo	Too small to be of commercial interest

Resources allows the public to join the search for NEAs of interest to scientists, space agencies and, of course potential asteroid mining companies. It is also designed to help to develop better computer-based search efforts that can more quickly train computer analysts and artificially intelligent programs to spot NEA motion against the fixed pattern of stars.

In time this project will likely expand to work from data obtained by the MIT Linear System, NEAT, Spacewatch, LONEOS, and other ground observatories. Efforts to generate new technology such as quantum computers might in time allow for a much more efficient means to search for the detection of previously unidentified NEAs.

Many of the NEAs detected in recent years have come from NASA's NEOWISE infrared telescope observations in space. The newly planned initiatives such as the B612 Foundation's Sentinel space telescope and NASA's NEOCAM will be able to detect much smaller NEAs that are closer at hand, since these devices having higher sensitivity and better orbits for such detection. There is sufficient capability now planned with more IR space telescopes, improved ground observation capabilities, and improved computer software for NEA detection to provide assurance that a number of suitable candidates for future space mining can and will be identified.

The current efforts by emerging space mining enterprises seem to be aimed at identifying candidates for mining that meet the following criteria: (1) reasonably large size, i.e., over 100 m in diameter; (2) approachable orbits that do not require additional velocity acceleration from low Earth orbit much in excess of 5 km/s, can be reached in under a year and also have a so called mission cadence (i.e., orbit that would allow start to finish return of valuable resources to Earth in less than around 4–5 years) and (3) a high percentage of valuable resources such as valuable metals (i.e., platinum) or water.

Not a large number of asteroids would be needed to make such an operation viable. If one were to confirm that an NEA such as 2011 CG2 was composed of something like 1 % platinum this would translate into something like 60,000,000 kg of platinum. The current market value of platinum is $40,000 per kg. Such a massive supply of platinum would obviously affect prices, but at current values this would translate into $2.4 trillion. Actually what is more likely is that the asteroid might consist of 1 % of so-called PGMs (or platinum-group metals). This would be a combined group of metals that includes platinum, palladium rhodium, osidium ruthenium and iridium.

In terms of technology needed to identify candidates for such mining operations, all of the key elements either seem to be in place, or will be in place well before the mining operations are actually begun for real. There is one interesting initiative that has begun under crowd-sourced "Kickstarter" funding by Planetary Resources, and this is the so-called Arkyd platform, which contains a small telescope that can be sent as a "scout" to review the viability of candidate NEAs in terms of size, mineral and metal content, etc. A test satellite named Arkyd 3 Reflight (A3R) was launched and successfully transported to Earth orbit on April 17, 2015, and was deployed from the International Space Station via the NanoRacks cubesat deployer on July 16, 2015.

Deep Space Industries has similar ambitions and plans to that of Planetary Resources. Their concept is also to launch probes similar to Arkyd spacecraft platforms. These probes would seek to examine possible candidate asteroids for mining in space. These smaller explorer craft are currently based on cost-effective cubesat technology and include "FireFly" (a 25-kg satellite) and "Dragon Fly." The plan is for these small probes to be launched as ancillary missions to much larger communications satellite launches. In December 2014, DSI also announced plans for a larger 150-kg "mothership" spacecraft that would be designed to deliver up to a dozen nanosats to trajectories that are beyond Earth orbit, for exploring distant asteroids and possibly other applications.[6] The problem with such ventures is that there is quite a bit of up front capital needed and much uncertainty for this type of high risk space venture, and thus obtaining the funds and long-term commitments to such ventures remain a real challenge. As yet, none of these probes has flown in space.[7]

For years interplanetary probes designed by space agencies were very complex and massive, but today probes with considerable scientific capability can involve payloads of a only a few kilograms.[8] As noted above Planetary Resources and Deep Space Industries are already developing exploration probes of the cubesat size and mass, such as the Arkyd platform and the Dragon Fly or even having a "mothership" where a single spacecraft might include something like ten micro-probes.

Thus, in terms of transportation technology, the mining of asteroids does not require any breakthrough development, as least for the exploration stage. Low cost chemical rockets can easily provide lift to low Earth orbit (LEO). After the launch to LEO either chemical or electrical ion propulsion systems can be used to achieve the 4.0–5.5 km/s increased velocity required to send quite small cubesat-type probes out to inspect asteroids to see if they are large enough and rich enough in metals and minerals and water to move to actual space mining operations. The larger technological question would involve how the "mined ore" would be returned to Earth in a safe and cost effective way.

This transport of mined ore to Earth, in terms of a shorter term proposition, probably most likely involves the issue of mining the Moon for minerals. Conventional solid rocket or liquid-fueled chemical rockets could lift robotic mining equipment to the Moon, but several new and "unconventional" transport means could be used to return the results of mining to the Earth. One of these could indeed become a two way system that could lift robotic mining material and other payload to the Moon and also provide for material to be returned to Earth as well.

New and Improved Transportation Technologies to Support Space Mining

Chemical Fueled Launchers

Chemical fueled rockets—both solid and liquid fueled systems—could support space mining operations, but these involve expensive consumables, and most systems today involve expendable launcher systems. New systems are coming on line all the time. These new systems include: Space X's Falcon 9 and Falcon 9 heavy, upgraded Indian and Chinese launch systems, Launcher One, the heavy lift Stratolaunch upper altitude launch system, Russian

[6] Woo, Marcus (2014-12-20). "Designing a Mothership to Deliver Swarms of Spacecraft to Asteroids," *Wired*. Retrieved 2014-12-17.

[7] Boyle, Alan (January 22, 2012). "Deep Space Industries' lofty asteroid ambitions face high financial hurdles," Cosmic Log. National Broadcasting Corporation. Retrieved January 23, 2013.

[8] "The 'CAPEd' Crusader: Goddard Technologist Advances CubeSat Concept for Planetary Exploration" *Satnews Daily*, May 21, 2015. http://www.satnews.com/story.php?number=71626023.

Angara, Ariane 6, Japanese H-3, Orbital Taurus II and ATK's Liberty. For lower mass asteroid reconnaissance missions there are also systems that can launch 200 kg and higher missions to LEO such as Launcher One and S-3 spaceplane systems All of these systems and more are all showing ongoing progress toward improved launcher systems that might support space mining operations with new launcher economies. (See Joseph N. Pelton's and Peter Marshall's, *Launching into Commercial Space* (2015), AIAA, Reston, Virginia.)

Space X, Blue Origin, S-3 and Stratolancher, in particular, are seeking to develop commercially efficient reuseable launcher systems using conventional chemical propulsion launchers. There is reasonable hope that within a 5-year time horizon such commercially developed chemically fueled vehicles will allow further significant launch cost reduction. It would also be hoped that several of these systems can also produce fewer pollutants The number one concern in this regard are particulates from such systems as Launcher One or the ATK Liberty that tend to use some of the dirtiest solid fuels, such as hydroxyl-terminated polybutadiene (HTPB) and aluminum polyamide. Particulates from solid-fueled rockets thus are of particular concern in terms of upper altitude pollution.

Ion Propulsion

The longer term future for space propulsion systems operating from the Moon, asteroids, and low gravity environments may be heavily influenced by electric or ion propulsion systems. Chemical thrust systems provide a great deal of high impulse thrust for short periods of time such as are needed to boost payloads into LEO. In contrast, electrical propulsion provides a greater amount of total impulse but only over a long period of operation. This produces a higher level of impulse thrust per kg of fuel over the entire period of operation.

Thus small, low mass surveyor probes that are launched to LEO by chemical rockets could then be spiraled out by low thrust ion propulsion systems to eventually reach a desired rendezvous orbit with targeted candidate asteroids. Today ion propulsion systems are primarily being developed for stabilization of large telecommunications or other application satellites. As this electrical ion thruster technology evolves it could find applications to move modules from the Moon, from asteroid mining operations, or to send small probes to candidate asteroids for possible future mining operations.

Ion thrusters operate through the use of either an electrostatic or electromagnetic force. Electrostatic ion thrusters use the so-called Coulomb force. This means that the ions accelerate to very high velocities in the direction of the electric field that is created within the electrical grid created within thruster device. There are alternatively what are called electromagnetic ion thrusters that employ the Lorentz force to accelerate the ions within what are commonly called plasma thrusters. These plasma thrusters do not typically use high voltage grids or positively charged anodes and negatively charged cathodes to accelerate the charged particles within the plasma. The electromagnetic ion thruster uses currents and potentials generated internally in the plasma that in turn serves to accelerate the plasma ions. This approach results in lower exhaust velocities due to the lack of high accelerating voltages but nevertheless has the advantage of longer life for the thruster.

The lack of high voltage grids that are found in electrostatic ion thrusters removes the problem of grid ion erosion over time. The plasma exhaust due to the lack of an electrical grid can be "quasi- neutral," which means that ion and electrons exist in equal number. This results in ion-electron recombination in the exhaust to neutralize the exhaust plume and eliminates the need for an electron gun or a hollow cathode. Also, the electrostatic ion thruster is limited to the use of noble gases, which is most typically xenon. The plasma thruster allows a much broader range of propellants that can include argon, carbon dioxide and other gases and liquids. The plasma ion thrusters, because of the elimination of the electrical grid that can erode under the plasma stream, are perhaps best suited for longer distance missions. In either case, with the electrostatic ion or pure plasma thruster, the thrust is created from the ion's kinetic energy. Still other ion thrusters utilize radio waves rather than an electrical field grid to generate the plasma that accelerates spacecraft.[9]

Ion thrusters as of the year 2015 typically have an input power spanning the 1–7 kW range of power. These thrusters also typically produce exhaust velocities that range from 20 to 50 km/s, with a thrust level that can be as high as 250 millinewtons and with a remarkably high efficiency of up to 80 %.[10]

The *Deep Space 1* spacecraft as designed by JPL and powered by an ion thruster as pictured below, was able to generate a delta-velocity increase of over nearly 4.5 km/s and did so by consuming less than 75 kg of xenon fuel. This is a much higher efficiency than a chemically fueled rocket, but the low thrust levels required an extended period of time to achieve the velocity increase (Fig. 3.2).

The Dawn spacecraft that was powered by an ion thruster has the current record of achieving a delta-velocity increase of some 10 km per second. This type of delta v is much greater than any candidate NEA would require that has been identified as a possible asteroid mining target. Considerable progress has been made in the United States, Russia, Europe, (Unfinished sentence?)

Nuclear Fueled Propulsion

There are a number of spacecraft that have derived power from a nuclear power source. These have usually involved long duration missions with relatively high power level requirements. Nuclear generators called (SNAP generators) and nuclear isotope power sources have been used for many years in cases where conventional solar and battery power systems were inadequate. More recently, however, there have been efforts to develop nuclear powered thrusters for various types of possible applications.

Nuclear fissile material as a power source for rocket propulsion presents

[9] Choueiri, Edgar Y "New dawn of electric rocket". *Scientific American* (2009). Issue 300: pp. 58-65.

[10] *Ibid.*

Fig. 3.2 The xenon ion thruster used on the *Deep Space 1* spacecraft (Graphic courtesy of NASA.)

lots of options. The least ambitious are nuclear isotope powered low level thrusters that might serve as the power source for ion thrusters. Such nuclear powered thrusters have been proposed for long-term maintenance of the International Space Station in LEO.

Other more ambitious options include a nuclear thermal rocket (i.e., the energy is used to heat the liquid hydrogen propellant on a larger scale), direct nuclear (fission products from a nuclear reaction directly propel the rocket), nuclear pulse propulsion (nuclear explosions propel the rocket), or, in the longer term future, some form of actual nuclear fusion could be developed. The recent breakthroughs in nuclear propulsion by the Lockheed Martin Skunkworks suggest that such options might become viable in a much shorter time horizon (i.e., another decade) than previously thought possible. Under such a scenario it might be a rocket directly powered by a small fusion reactor or perhaps it might be possible to use heat from a fusion reactor to provide the continuous power source for a rocket propulsion system or perhaps to power a lunar colony or remote space mining operation.

Helium-3 propulsion would use the fusion of helium-3 atoms as a power source for spacecraft. An isotope of helium with two protons and one neutron, helium-3 could be fused with deuterium in a reactor. The resulting energy release could be used to expel propellant out the back of the spacecraft. In recent years helium-3 has been proposed as a power source for spacecraft mainly because of its abundance on the Moon. Currently, scientists estimate that there are one million tons of helium-3 present on the Moon. This buildup of helium 3 is thought to be mainly due to the solar wind, containing helium-3, colliding with the Moon's surface and depositing it, among other elements, into the soil.

Others are thinking more near-term. Shackleton Energy has announced its plans, not so much the mining of helium-3 to provide fuel for nuclear fusion but rather simply mining for water to create hydrogen and oxygen for chemical propulsion. The Shackleton Energy website opens with the following declaration: "We Are Going Back to the Moon to Get Water. There are billions of tons of water ice on the poles of the Moon. We are going to extract it, turn it into rocket fuel and create fuel stations in Earth's orbit. Just like on Earth you won't get far on a single tank of gas, what we can do in space today is straight-jacketed by how much fuel we can bring along from the Earth's surface. Our fuel stations will change how we do business in space and jump-start a multi-trillion dollar industry. Much like gold opened the West, lunar water will open space like never before."[11]

The problem with many of these concepts is that whether one is mining the Moon for helium-3 or frozen water, the Space Treaty of 1965 declared that the "Moon and Other Celestial Bodies" were a part of the global commons, and it certainly did not allow for the removal of resources from the Moon. The Gold Rush of 1849 was carried out inside the continental United States, but the Moon is not so easily exploited under existing international law.

Mass-Driver Systems on the Moon

There are many other options that might provide key new transportation capabilities for space mining that are easier and safer to use to support space mining. One of the most discussed and analyzed is the idea called "mass drivers," which was popularized by Dr. Gerard K. O'Neill, author of the well-known book *The High Frontier*.

O'Neill's concept was to create a mass driver on the surface of the Moon that could launch mass off the surface at rates such as 2 kg every 30 s or 240 kg an hour, or about 3 metric tons a day.[12] This would clearly be something that could also be accomplished on a large asteroid, assuming that sufficient power was available to support this type of transport. Because the gravitational pull of an asteroid would be much less than that of the Moon, such a mass driver would be able to send a much higher volume into space from the NEA. The key question is what would be the collection point for such a stream of mined ore from either the Moon or an asteroid.

Space Elevator Systems

If one thinks even longer term there have been various proposals and studies of a possible space elevator from Earth

---

[11] Shackleton Energy Overview: http://www.shackletonenergy.com/overview/#goingbacktothemoon.

[12] Gerard O'Neill, *The High Frontier: Human Colonies in Space*, (2000) Apogee Books, Burlington, Canada.

to geostationary orbit and even a cable transportation system between Earth and the Moon. The design of such a space elevator or space funicular with solar-powered robotic climbers, a cable of sufficient tensile strength and radiation hardening is currently beyond human engineering capabilities and may in the end not to be economically viable to build, deploy and operate. If such systems ultimately do prove to be possible, though, this could provide a significant breakthrough for higher volume space mining activities.[13]

Conclusions

The technologies associated with space mining continue to improve. Many new ideas with regard to space transportation, space power systems, and robotic mining systems have evolved in the last 20 years, since the time John Lewis wrote one of the more popular books on this subject entitled *Mining the Sky*. Certainly we have come much further than 60 years ago, when science fiction thrillers such as *Rip Foster Rides the Gray Planet* were first inspiring young entrepreneurs to think that space mining might indeed be possible. [14]

Despite significant technological progress on essential engineering skills, and also a much greater need due to depletion of Earth's vital ores, there remains much more to be done. This chapter only provides an introduction to some of the key transportation systems that will need to be developed further to make space mining possible—when and if the legal, regulatory and standards issues can be resolved to make such enterprises possible.

[13] Bradley C. Edwards and Eric Westling, *The Space Elevator* (2003) Praxis Books, NY.

[14] John Lewis, *Mining the Sky: Untold Riches from the Asteroids, Comets and Planets* (1997) Perseus books, NY.

4 Power and Robotic Systems for Space Mining Operations

The essential elements for space mining operations include robotic systems that can carry out the mining operations, a sustaining power source that can allow the robotic systems to operate, the sensing ability to locate valuable ores, water or minerals on the Moon or asteroids for possible recovery, and transportation systems to and from the sites for the mining operations. The previous chapter discussed the process of sending surveyor probes out to identify likely sources for space mining and transportation systems, while this chapter discusses the robotic mining and sustaining power systems. This is not to say that there are no other technologies and systems that might need to be developed, but these are the primary technical needs. It is possible that in the future 3D printing systems armed with the right software and raw materials could implement the infrastructure for an entirely functional space colony or other sophisticated tasks, but for near-term space mining activities, advanced survey and prospecting, transportation systems, robotic mining systems and sustaining power systems represent the true core technologies.

Any technical aspects and the design, manufacture, and operation of technological systems could be subject to legal, regulatory, or standards at a future date. Some of these constraints, for instance, might relate to pollution, i.e., the use of nuclear or radioactive materials or particulates from solid fuel rockets or thrusters; to safety standards, i.e., the use of nuclear or radioactive materials, or intellectual property rights; or to the proper use of patented inventions. For the most part these technological systems do not seem to raise major legal, regulatory or standards issues that could not be fairly easily resolved in a national court and legal justice system, but clearly become complicated if these operations are carried out on the Moon or asteroids, where no national laws, regulations or safety standards are in affect. The precedents established with regard to Antarctica and national research and exploration stations are perhaps the most useful guidance that might apply to space mining of the Moon or asteroids.

Clearly the "elephant in the room" is whether space mining by a private entity, consortium, or governmental body for its own advantage—rather than

© Springer International Publishing Switzerland 2017
R.S. Jakhu et al., *Space Mining and Its Regulation*,
Astronautical Engineering, DOI 10.1007/978-3-319-39246-2_4

on behalf of the "global commons"—is allowable. This is the fundamental issue, and the choice of technologies for space mining operations is secondary to this key issue. All these legal and regulatory issues are considered in later chapters.

Power Systems

In order to support the telecommunications and data networking needs to the robotic mining equipment, local transportation of the mined materials, and the actual mining operations, there clearly must be sustainable power systems. Fortunately there are a number of possible options.

Photovoltaic and Battery Systems

Some time ago solar (photovoltaic) cells were developed that can generate electrical power from the Sun's radiation, and battery systems were developed for times when solar illumination is blocked. This is clearly an available technology that is well proven. Higher efficiency photovoltaics, such as so-called "multi-junction violet" cells, gallium arsenide cells[1] or quantum dot technology are being developed and can provide higher efficiency power conversion.[2] Lithium ion batteries can also support higher energy storage densities and longer life.

The solar-cell-plus-battery-power system makes sense in that no fuel needs to be provided to operate such systems. It is likely that solar energy will the prime power source to support telecommunications, data networking, and initial space mining operations. Fairly large solar arrays, however, will likely be needed to support drilling operations. It is thus likely that the arrays will also be designed with solar concentrators so that the photovoltaics will be able to absorb the equivalent of several Suns. Likewise there is likely to be a tracking system so that maximum illumination from the Sun will be constantly available. There will also likely be some glass coating on the solar cells to maintain and prolong their lifetime.

Even so, if the mining operation were to be sustained for a period of longer than 15 years, it might be necessary to replace the worn out solar cells and batteries. It is possible to design the solar cell power system in the first place so as to provide replacement components that could be robotically installed as needed.

Space-Based Thermocouple Energy Systems

Another possibility might be the use of as space-based thermocouple. Neil Ruzik patented the idea of using the extreme temperature gradients on the Moon between the Sun's illuminated side and the dark side. The principle of a thermocouple involves two plates, each made of a different metal that conducts electricity. Joining these two plates to form a closed electrical circuit while keeping the two junctions at different

[1] "Silicon vs. Gallium Arsenide Which Photovoltaic Material Performs Best," NASA Tech Briefs, January 1, 2014. http://www.techbriefs.com/component/content/article/27-ntb/features/application-briefs/18946.

[2] Quantum Dot Solar Cells Break Conversion Efficiency Record http://spectrum.ieee.org/nanoclast/green-tech/solar/quantum-dot-solar-cells-break-conversion-efficiency-record.

temperatures produces an electric current. Each pair of junctions can be used to form an individual thermocouple. Thus it would be possible to create a large number of thermocouples across the sunlight "discriminator" to generate a significant amount of electricity. It would be possible to use a solar lens concentrator to heat one end of a space-based thermocouple and place the other end on the cold side of an asteroid or the Moon to generate electricity. The thermocouple would not likely be as efficient as solar cells, but it could have a longer life and be less expensive to build and install.[3]

Nuclear or Radioactive Isotope Power System

A nuclear or radioactive isotope power system could provide a reliable longer-term source of power that would not require solar illumination. For a mining operation where extensive deep drilling to obtain the ores was needed, this use of nuclear energy might be considered as an option from the outset. Radioisotope thermoelectric generators, or RTGs, provide electrical power for spacecraft by converting the heat generated by the decay of an isotope. In the case of the current RTG the radioisotope power source is plutonium-238 (pu-238). Electricity is generated via thermocouples as discussed above. Since there are no moving parts that could wear out, RTGs have historically been viewed as a highly reliable power option. Thermocouples have been used in RTGs by NASA for space exploration missions for a total combined time of over 300 years, and within each of nearly 20 RTGs there has never been a failure of the thermocouple (see Fig. 4.1).

Within the RTG, the radioisotope fuel heats one of these junctions while the other junction remains unheated and is cooled by the space environment or a planetary atmosphere.

The current RTG model in use by NASA is the so-called multi-mission radioisotope thermoelectric generator, or MMRTG. The design is closely based on the type of RTG flown previously on the two Viking landers and *Pioneers 10* and *11*.

The MMRTG produces electricity at an operational efficiency of only 6–7 %. Multiple MMRTGs could be combined to provide higher levels of electrical power for a mining mission. Each MMRTG contains just under 5 kg of plutonium-238 dioxide as its nuclear fuel, using eight general purpose heat source (GPHS) modules to produce about 110 W of electrical power in total. Thus five such units would be needed to produce 550 W of electrical output. The excess heat from an RTG could be used to maintain the temperature of a robotic drilling and processing unit at a desired operating thermal level. There is the additional complication that such RTG power systems are subject to close governmental supervision and control in relation to who might legally use them. In addition there are separate and quite strict controls on who and under what regulatory authority can launch such systems into space. A commercial operator might find it difficult to obtain licenses to own and operate such a power source and even more difficult to get approval to launch one or more of these RTG's into orbit.

[3] Thermocouple in outer space: www.physics-edu.org/tech/thermo_electricity_in_outer_space.htm.

Fig. 4.1 The radioisotope thermal-electric generator now in use for the Curiosity rover on Mars (Image courtesy of NASA.)

It is possible that a space mining operation, for the purposes of flexibility, reliability, and efficiency, might conceivably want to combine all three types of power sources, namely solar cell/batteries, a space-based solar discriminatory thermocouple, and a radioisotope thermal generator.[4]

Thermoionic Power Source

There is at least one more possible power source technology that might be of use to space mining operations. This is quite similar to a thermocouple device in terms of how electricity is generated, except in this case both light and/or heat could be converted into electricity. In the latest designs an electrical grid is inserted in the gap between the heated and non-heated metal plates.

This approach has been known for some time, but it operates at a low efficiency. Only recently researchers at the Max Planck Institute for Solid State Research in Stuttgart, the University of Augsburg and at Stanford University have come up with a new way of creating an electric field in the space between the plates to solve what is called the space-charge problem. This field works to accelerate the electrons leaving the hot plate and then again slows them down as they approach the cold plate. This thus allows a continuous flow of current to be established. The field that allows this continuous flow is created by a honeycomb-patterned gate with hexagonal holes in it. The electrons are guided through the holes by applying a magnetic field between the plates. Although current systems using this

[4] Radioisotope power system: https://solarsystem.nasa.gov/rps/rtg.cfm.

design are only about 10 % efficient it is now thought that up to 40 % efficiency might ultimately be achieved.[5]

Explosives as a Substitute for Mining-Related Energy Needs

Just as is the case of conventional mining, it is of course possible to use conventional explosives to assist with mining operations. There are significant special issues, though, that make explosives difficult to use in the low gravitational environment of outer space. A whole new field of research will be needed to devise shaped explosions within protected or shielded explosion containment areas so as to recover minerals or metals without indiscriminately blasting materials into space.

By way of example, the orbital escape speed for the moons of Mars, Phobos and Deimos, is just about 50 km/h, or about 14 m/s. Any uncontrolled explosion for an asteroid would typically send all of the fragments off into space with no easy way to recover them, unless the entire asteroid was contained within a very strong netting.

Space Robotic Mining Systems

As noted earlier in this book, many key metals and rare earth minerals are now depleted from Earth's easily accessible reserves. In short, large, easy to get at deposits have all but disappeared from Planet Earth. The resources of tomorrow are now thought to be located in most remote parts of the globe, many hundreds of meters beneath the surface, or perhaps on the Moon or on NEAs. Robotic mining equipment developed for extremely deep excavation could provide prototype concepts for robotic mining in space. The Osiris mission by NASA is intended to provide new information about practical space mining operations on an asteroid (see Fig. 4.2).

There have already been several detailed designs for engineering devices that could be used in developing a robotic mining system for NEAs. The graphic below provides just one illustration of such a device.[6]

Serious studies of the challenges to be overcome have identified a number of potential problems. At a high level of abstraction the challenges include the following:

1. Finding ways to conduct mining with lower levels of energy consumption.
2. Minimizing the shape, size and mass of robotic mining equipment.
3. Improved software and artificial intelligence to allow tele-operation and automation.
4. Reduction of complexity, improved ease of operation, and reduced need for lubrication and machinery that needs frequent maintenance and human attention.

[5] http://physicsworld.com/cws/article/news/2013/dec/09/new-generator-creates-electricity-directly-from-heat.

[6] NeoMiner—Robotic Asteroid Mining Equipment used for extracting metals from Near Earth Objects and further distant asteroids Asteroid Mining Equipment https://www.google.com/search?q=robotic+mining+equipment.

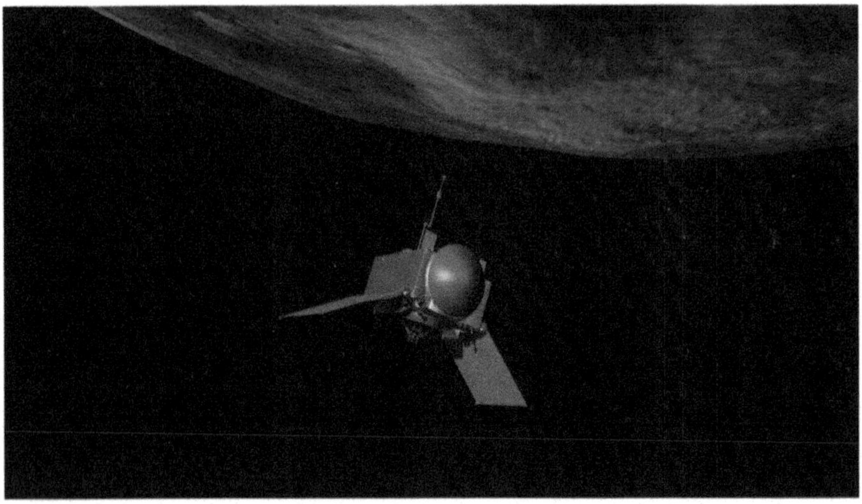

Fig. 4.2 NASA Osiris spacecraft depicted in space (Graphic courtesy of NASA.)

5. Ease of assembly or fabrication in space via "smart" 3D printers.
6. Possible manufacture from locally acquired materials (see 5 above).
7. High level of durability and reduce the need for maintenance and repair (see 4 above).
8. Limited need for resupply of any consumables.
9. Compatibility with local gravity, vacuum, solar illumination, radiation, dust generated by mining and thermal environment. (Coping with the harsh space environment, for example, with no human repair or maintenance personnel on hand is a huge challenge.)
10. Upgradability of systems and modularity of design so that improved technology or broken or obsolete parts can be replaced.

These challenges are so basic that it is clear that this is a much more difficult design and engineering feat to achieve than simply adapting equipment developed for deep Earth robotic mining. In short, the equipment that has been developed for Earth-based operations is too large, too massive, too complex, too energy intensive, too geared to a one-g environment, and too in need of frequent maintenance and repair to be used effectively in space.[7]

Innovative New Space Mining Concepts

The challenges associated with robotic space mining equipment is such a large challenge, it may well be that totally new techniques that might be uniquely applied to NEA mining might be essential to future progress in this area. Nikola Tesla speculated that high-powered

[7] Peter Chamberlain, Lawrence Taylor, Egons Podnieks, and Russell Miller, "A Review of Possible Mining Applications in Space," University of Arizona Press. http://www.uapress.arizona.edu/onlinebks/ResourcesNearEarthSpace/resources03.pdf

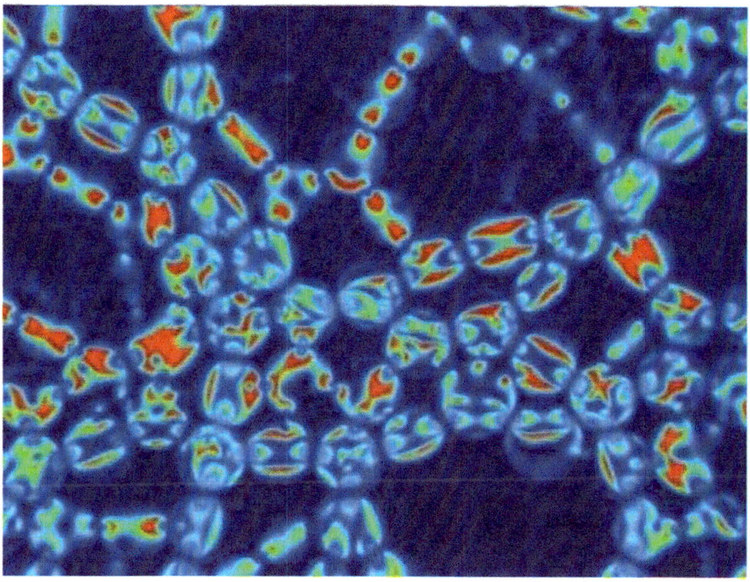

Fig. 4.3 Force chains (shown in *red*) that reveal fault lines subject to sonic intensification (Image courtesy of U. S. Los Alamos Department of Energy Laboratory.)

electromagnetic weapons systems might be developed in the future. Techniques that are potentially dangerous if applied on Earth might be applied to mining operations on NEAs to create chunks of asteroids. It is possible that once the "space rocks" are more reasonably sized, they could then be returned to Earth or put in lunar orbit. At this time, however, current plans by space mining companies are focused on what would be called conventional mining techniques.[8]

Researchers as the U. S. Los Alamos laboratories have reported on their research in this area as follows: "Using a novel device that simulates earthquakes in a laboratory setting, a Los Alamos researcher and his colleagues have shown that seismic waves—the sounds radiated from earthquakes—can induce earthquake aftershocks, often long after a quake has subsided. The research provides insight into how earthquakes may be triggered and how they recur."[9] (see Fig. 4.3)

It also might be noted that an NEA or a sizable chuck of an asteroid could be put into Earth or lunar orbit or perhaps L-1 (Lagrange point) as a potential protective device to be positioned defensively against the impact of a city-killing asteroid on a path to bring destruction to Earth.

[8] "Listening to the 9.0-magnitude Japanese earthquake: Seismic waves converted to audio to study quake"s traits", *Science News*. http://www.sciencedaily.com/releases/2012/03/120306142506.htm.

[9] "Sound Waves Can Trigger Earthquake Aftershocks" Research report from the U.S. Department of Energy Los Alamos Laboratories http://www.sciencedaily.com/releases/2008/01/080103124649.htm.

The development of these new techniques and capabilities involve more than technical innovation, and could raise significant issues of international space law. The development of large amplitude sonic frequency generators for space mining systems might also be considered the equivalent of developing a space weapon. Further, the deployment of a new asteroid or parts of asteroids into any new near-Earth orbit also leads to new process issues as well. There would be questions as to whether the movement of such "significant space objects" either into or out of Earth orbit would need to be approved by the United Nations' sanctioned Space Mission Planning Group or some other international approval process.

Conclusions

The technical challenges of space mining, either on the Moon or on NEAs, are considerable. The development of a consistent and sufficient power supply and the development of mining equipment that can overcome the ten major constraints identified earlier are truly epic. It may very well be that very different and innovative techniques would need to be developed to make off-Earth mining viable. These new approaches and redeployment of "significant space objects" could well present not only new technical but also new legal and regulatory challenges as well.

5 U. S. Space Exploration and Planetary Resources

The United States has been carrying out a vigorous space exploration program since the inception of NASA nearly 60 years ago. This has included space telescopes, scientific sensing and data collection missions. It has included orbiters, landers and rovers. NASA has also supported the development of space applications in communications, remote sensing, and navigation, and has most recently undertaken prize competitions related to space mining technologies and commercial space transportation development. NASA has indeed awarded developmental and research contracts to some of the commercial space mining ventures discussed in the previous chapter. Because NASA has carried out many hundreds of missions over its existence it is not possible to cover every one of its activities. This chapter, however, seeks to highlight the most relevant of its planetary and resource exploration missions over time.

Space Telescope Missions

NASA launched the Hubble Space Telescope observatory in April 1990 to observe stars, planets and galaxies. Over time the Hubble has provided an enormous amount of useful information about the Moon, asteroids and the planets and their moons in addition to its deep space observations.[1] The Hubble Space Telescope is international in scope, and although the program has been led by NASA there has been substantial scientific involvement by the European Space Agency (ESA).[2]

NASA launched the Kepler spacecraft on March 2009. The primary objective of the Kepler mission was "to survey .... [a specifically targeted portion] of the Milky Way galaxy to discover hundreds of Earth-size and smaller planets in or near the habitable zone and determine the fraction of the hundreds of billions of stars in our galaxy that might have such planets." The scientific objective of the mission continues to be "to explore the structure and diversity of planetary systems by

[1] Hubble Space Telescope: NASA.gov http://www.nasa.gov/mission_pages/hubble/story/index.html#.VOE0yfnF-Sq (Accessed August 27, 2015)

[2] Hubble Overview, Space Science, ESA http://www.esa.int/Our_Activities/Space_Science/Hubble_overview (Accessed August 27, 2015)

© Springer International Publishing Switzerland 2017
R.S. Jakhu et al., *Space Mining and Its Regulation*, Astronautical Engineering, DOI 10.1007/978-3-319-39246-2_5

surveying a large sample of stars to ..... [d]etermine the properties of those stars that harbor planetary systems."[3]

NASA, in collaboration with ESA and the Canadian Space Agency [CSA], is currently scheduled to launch the James Webb Space Telescope in October 2018. This exceptionally large and sophisticated telescope "will be the premier observatory of the next decade" and "will study every phase in the history of our Universe, ranging from the first luminous glows after the Big Bang, to the formation of solar systems capable of supporting life on planets like Earth, to the evolution of our own Solar System."[4]

NASA has selected the Transiting Exoplanet Survey Satellite (TESS) mission for launch in 2017 to "find exoplanets transiting nearby, bright stars." The TESS mission is considered the best way to identify targets for follow-up characterization with large ground telescopes, the Hubble Telescope, and the Webb Telescope.[5]

The Apollo Lunar Exploration Program

The Apollo program was originally launched with the specific objective to land Americans on the Moon and returning them safely to Earth. With success of the *Apollo 11* mission that met the initial objective, the goals of the Apollo program were broadened to include carrying out scientific exploration of the Moon. As part of this exploration, astronauts collected various samples from the Moon through the last mission of the program, the *Apollo 17* mission, in December 1972. The extensive sample collection from the Apollo program serves as one of the most extensive databases concerning the chemical composition of the Moon.[6]

The Mariners, MESSENGER, the Voyagers, Galileo, the Pioneers, Juno, Huygens-Cassini, Magellan and New Horizons

NASA's Mariner series of missions were planned to be the "first U. S. spacecraft to other planets, specifically Venus and Mars." The successful Mariner missions were: *Mariner 2* (launched August 1962 to Venus), *Mariner 4* (launched November 1964 to Mars), *Mariner 5* (launched June 1967 to Venus), *Mariner 6* (launched February 1969 to Mars), *Mariner 7* (launched March 1969 to Mars), *Mariner 9* (launched May 1971 to Mars), and *Mariner 10* (launched November 1973 to Venus and Mercury).[7]

The mission objective of *Mariner 2* was to study Venus and to make close up scientific observations. The *Mariner 4*

[3] Kepler Overview, Kepler:NASA.gov http://www.nasa.gov/mission_pages/kepler/overview/index.html#.VOkYSvnF-So (Accessed August 27, 2015).

[4] About the James Webb Space Telescope, Explore James Webb Space Telescope, NASA http://jwst.nasa.gov/about.html.

[5] NASA Facts: TESS: Transiting Exoplanet Survey Satellite, FS-2014-1-120-GSFC (October 2014), online: TESS, NASA http://tess.gsfc.nasa.gov/documents/TESS_FactSheet_Oct2014.pdf

[6] Apollo Program, NASA http://www.nasa.gov/mission_pages/apollo/missions/index.html#.VM1iz2jF-Sr.

[7] Mariner Missions, NASA http://science1.nasa.gov/missions/mariner-missions (Accessed August 27, 2015).

mission was essentially to make parallel observations of the planet Mars: "to conduct [close-up] scientific observations of Mars." The more detailed objectives for the *Mariner 5* mission were to "collect data on [the Venusian] atmosphere, radiation and magnetic field." The chief objectives for *Mariners 6* and *7* were to make finer grain observations for Mars than had been possible with earlier Mars missions. The objectives of the *Mariner 9* mission were to map 70 % of the Martian surface and study "temporal changes in the Martian atmosphere and on the Martian surface." The primary scientific objectives of the *Mariner 10* mission were "to measure Mercury's environment, atmosphere, surface, and body characteristics and to make similar investigations of Venus."[8]

Another mission involving Mercury was NASA's MESSENGER spacecraft. The MESSENGER mission was launched on August 2004 with the objective of providing the first detailed observations of Mercury.[9] The latest European Space Agency exploratory mission to Mercury, "BepiColombo," to be carried out in cooperation with the Japanese Space agency (JAXA), is described in the chapter on European activities.

NASA launched its Voyager twin spacecraft mission to explore Jupiter and Saturn in the summer of 1977. *Voyager 1* was actually the second of the two and was launched on September 5, 1977, while *Voyager 2* had lift-off 2 weeks before on August 20, 1977. Originally, the mission was designed to study only Jupiter and Saturn. The two-spacecraft mission was targeted to conduct "close-up studies of Jupiter and Saturn, Saturn's rings, and the larger moons of the two planets." The mission was later renamed Voyager Neptune Interstellar Mission after NASA authorized a further Neptune leg of the Voyager mission.[10] Currently, the mission is known as the Voyager Interstellar Mission. Under this current mission, the two spacecraft have been further programmed to continue to explore ultraviolet sources among the stars, and the fields and particles instruments aboard the Voyagers will continue to explore the boundary between the Sun's influence and interstellar space.

NASA launched the *Galileo* spacecraft on October 18, 1989, to make a detailed study of Jupiter and its moons and magnetosphere from orbit. On September 21, 2003, "Galileo plunged into Jupiter's crushing atmosphere" and was destroyed. One of the most important discoveries of the Galileo mission before the spacecraft was destroyed was that there might possibly be an ocean beneath the icy crust of the moon

[8] Mariner 2, NASA http://www.jpl.nasa.gov/missions/mariner-2/.

Mariner 4, NASA http://nssdc.gsfc.nasa.gov/nmc/spacecraftDisplay.do?id=1964-077A.

Mariner 5, NASA http://www.jpl.nasa.gov/missions/mariner-5/.

Mariner 6 & 7, Mars Exploration, NASA http://mars.jpl.nasa.gov/programmissions/missions/past/mariner6/7/.

Mariner 9, NASA http://nssdc.gsfc.nasa.gov/nmc/spacecraftDisplay.do?id=1971-051A.

Mariner Missions, NASA http://science1.nasa.gov/missions/mariner-missions/ (All accessed on August 27, 2015)

[9] Launch Coverage, MESSENGER, NASA http://www.nasa.gov/mission_pages/messenger/launch/index.html.

[10] NASA Facts: Voyager to the Outer Planets and Into Interstellar Space, JPL 400-1538 09/13 (Accessed September 2013). http://www.jpl.nasa.gov/news/fact_sheets/voyager.pdf

Europa." The upcoming ESA mission to explore Jupiter and three of its moons of Ganymede, Europa and Callisto in 2022 will try to determine if possibly all three might have underground oceans.[11]

Even before the Galileo mission, NASA launched *Pioneer 10* and *Pioneer 11* spacecraft on March 1972 and April 1973, respectively, to fly to Jupiter, Saturn, and ultimately other parts of the Milky Way Galaxy. The science mission of *Pioneer 10* ended on March 1997 and the *Pioneer 11* mission ended on September 1995.[12]

Another mission of NASA involving Jupiter is the Juno mission that is currently underway. The Juno spacecraft was launched on August 2011, and the stated mission objective is to "improve our understanding of the Solar System's beginnings by revealing the origin and evolution of Jupiter." Juno is expected to enter Jupiter's orbit on July 2016. Specifically, Juno will carry out the following tasks:

- Determine how much water is in Jupiter's atmosphere, which would help to determine which planet formation theory is correct (or if new theories are needed).
- Look deep into Jupiter's atmosphere to measure composition, temperature, cloud motions and other properties.
- Map Jupiter's magnetic and gravity fields, and thus help to reveal the planet's deep structure.
- Explore and study Jupiter's magnetosphere near the planet's poles, especially the auroras—Jupiter's northern and southern lights—thus providing new insights about how the planet's enormous magnetic force field affects its atmosphere. If Juno survives long enough it might coordinate its measurements with the ESA JUICE project.[13]

NASA, jointly with the ESA and the Italian space agency, Agenzia Spaziale Italiana [ASI], launched the Cassini-Huygens mission in October 1997 to carry out in-depth studies of Saturn and its moons, rings and magnetic environment. Onboard the Cassini spacecraft was a scientific probe called Huygens that was released from the main spacecraft "to parachute through the atmosphere to the surface of Saturn's largest and perhaps most interesting moon, Titan." Cassini arrived at Saturn in 2004, and the ESA's Huygens probe was released on December 24, 2004, and landed on Titan's surface on January 14, 2005. Cassini's initial 4-year mission ended in June 2008 but continued with the "first extended mission, called Cassini Equinox Mission, in September 2010." The second extended mission, called the Cassini Solstice Mission, will continue until September 2017. It is hoped that this mission will help to determine "the physical state,

[11] Galileo, Solar System Exploration, NASA http://solarsystem.nasa.gov/missions/profile.cfm?Sort=Chron&StartYear=1980&EndYear=1989&MCode=Galileo

[12] "Pioneer-10 and Pioneer-11" (26 March 2007), online: Mission Archives, NASA http://www.nasa.gov/centers/ames/missions/archive/pioneer10-11.html Also see Pioneer, Program, Jet Propulsion Laboratory http://space.jpl.nasa.gov/msl/Programs/pioneer.html.

[13] "Unlocking Jupiter's Secrets" (24 August 2011), online: Overview, Juno, NASA http://www.nasa.gov/mission_pages/juno/overview/index.html#.VOkGuPnF-So.

topography and composition of Titan's surface and characterize its internal structure."

This spacecraft's observations of the icy moons of Saturn are seeking to reveal "the compositions and distributions of surface materials," and also help reveal "the bulk compositions and internal structures of the moons."[14]

The Pioneer Venus Orbiter that is also known as the *Pioneer 12* mission was "designed to perform long-term observations of the Venusian atmosphere and surface features." This spacecraft was launched by NASA in May of 1978. *Pioneer 12* continued operation until October 1992. Pioneer Venus Multiprobe, or the *Pioneer 13* mission, was designed to perform detailed atmospheric measurements of Venus. It was launched on August 1978 and burned up in Venus's atmosphere on December 1978.[15]

In May 1989, NASA launched the Magellan spacecraft to develop an understanding of the geological structure of the planet Venus, including Venus's density distribution and dynamics. The mission ended on October 11, 1994, when the spacecraft was "commanded to plunge into [the Venusian] dense atmosphere."[16]

NASA launched the New Horizons mission on January 2006 to study for the first time the Pluto system and the Kuiper Belt. On July 14, 2015, the New Horizon spacecraft had its closest approach to Pluto and its moons. Pluto and its largest moon, Charon, belong to a third category of planetary bodies known as "ice dwarfs." They have solid surfaces but, unlike the terrestrial planets, a significant portion of their mass is icy material. Using images from the Hubble Space Telescope images, New Horizons team members have now discovered four previously unknown moons of Pluto: Nix, Hydra, Styx and Kerberos that are also composed of icy material.[17]

A close-up look at these worlds as produced from New Horizons data is expected to reveal more information about the origins of Pluto and objects on the outskirts of our Solar System and the Kuiper Belt, from which comets originate. New Horizons is exploring—for the first time—useful information about the origin of ice dwarf planets such as Pluto and Kuiper Belt bodies and how they may have evolved over time (see Fig. 5.1).

Lunar Missions

NASA launched the Clementine probe on January 25, 1994, which was a joint project between the Strategic Defense Initiative Organization and NASA. The objective of the Clementine mission was to "test sensors and spacecraft components under extended exposure to the space environment and to make scientific observations of the Moon and

[14] NASA, Press Kit: Cassini-Huygens Saturn Arrival (June 2004) http://www.nasa.gov/pdf/60116main_cassini-arrival.pdf.

[15] Pioneer 12, Quicklook, Jet Propulsion Laboratory http://space.jpl.nasa.gov/msl/QuickLooks/pioneer12QL.html.

[16] "Magellan Mission at a Glance," http://www2.jpl.nasa.gov/magellan/fact.html (Accessed August 27, 2015).

[17] New Horizons: The First Mission to the Pluto System and the Kuiper Belt (August 2014), online: NASA http://www.nasa.gov/sites/default/files/files/NHMissionFS082114HiPrint.pdf (Accessed Aug. 2014).

Fig. 5.1 The New Horizons spacecraft that flew by Pluto on July14, 2015 (Graphic courtesy of NASA.)

the near-Earth asteroid 1620 Geographos." The observations to be made from the Clementine were "originally for the purposes of assessing the surface mineralogy of the Moon and Geographos, obtaining lunar altimetry from 60N to 60S latitude, and determining the size, shape, rotational characteristics, surface properties, and cratering statistics of Geographos." Most significantly Clementine also confirmed a significant amount of water (ice) on the Moon and was able to complete its mission at a small cost compared with other NASA exploratory projects. Due to malfunction in one of the on-board computers on May 7, 1994, the planned continuation of the mission became impossible.[18]

[18]Clementine Project Information, NASA http://nssdc.gsfc.nasa.gov/planetary/clementine.html.

There are several other missions that concern the Moon, namely, the Lunar Prospector (LP) mission, the Lunar Reconnaissance Orbiter (LRO) mission, the Lunar CRater Observing and Sensing Satellite (LCROSS) mission and the Lunar Atmosphere and Dust Environment Explorer (LADEE) mission. The LP mission was launched in January 1998 "for a one-year, polar orbit, primary mission dedicated to globally mapping lunar resources, gravity, and magnetic fields, and even outgassing events." The LRO and LCROSS were launched together on June 2009. LRO's mission was "to map the moon's surface and, after a year of exploration was extended" to continue mapping. The principal mission objective of LCROSS mission, which is a companion mission to LRO, is "to confirm the presence or absence of water ice in a permanently shadowed crater near a lunar polar region." The CLROSS mission thus used information learned from the Clementine and LP missions. The LADEE mission was launched on September 2013 and impacted the lunar surface in April 2014. The object of the mission was to "orbit the moon in order to gather detailed information about the lunar atmosphere, conditions near the surface and environmental influences on lunar dust."[19]

Mars Missions

NASA has devoted a tremendous amount of resources to the study of MARS going back over 40 years. The following are highlights from the various scientific and exploratory missions.

NASA's first successful mission to land on Mars was its Viking mission. Launched in August 1975, the mission was composed of two pairs of spacecraft, namely *Viking 1* and *Viking 2*, and was "designed to take high-resolution images of the Martian surface, characterize the structure and composition of the atmosphere and surface, and search for evidence of life." With some careful planning this spacecraft alighted on Mars on July 4, 1976, to celebrate the 200th Independence Day in the United States.[20]

NASA launched the Mars Global Surveyor mission, a global mapping mission, in November 1996, which examined the entire planet. The mission "contributed to meeting the four main science goals of the Mars Exploration Program," namely, determine whether life ever arose on Mars, characterize the Martian climate, characterize the Martian geology, and prepare for human exploration. The spacecraft went silent on November 2006.[21]

NASA's second mission under its Discovery Program is the Mars Pathfinder mission. which was launched in December 1996 and ended in September 1998. The mission was "designed to be a demonstration of the technology necessary to deliver a lander and a free-ranging robotic rover to the surface of Mars in a cost-effective and efficient manner."[22]

In April 2001, NASA launched the 2001 Mars Odyssey mission, which is a part of NASA's Mars exploration program. To contribute to the four main science goals of the Mars exploration program, noted above, and meet the mission's specific goal of determining the habitability of Mars, Odyssey has five science objectives: globally map the elemental composition of the surface, determine the abundance of hydrogen in the shallow subsurface, acquire high

[19] Lunar Prospector Mission, LCROSS, NASA http://www.nasa.gov/mission_pages/LCROSS/searchforwater/lunar_prospector.html.

LRO Overview, Lunar Reconnaissance Orbiter, NASA http://www.nasa.gov/mission_pages/LRO/overview/index.html.

LCROSS Overview, LCROSS, NASA http://www.nasa.gov/mission_pages/LCROSS/overview/index.html.

LADEE, NASA http://www.nasa.gov/mission_pages/ladee/main/index.html and http://www.nasa.gov/mission_pages/ladee/launch/index.html#.VOkwr_nF-So.

[20] Viking 01, Solar System Exploration, NASA http://solarsystem.nasa.gov/missions/profile.cfm?Sort=Alpha&Letter=V&Alias=Viking%2001.

[21] Overview, Mars Global Surveyor, NASA http://mars.jpl.nasa.gov/mgs/overview/.

[22] Overview, Mars Pathfinder, NASA http://www.nasa.gov/mission_pages/mars-pathfinder/index.html.

spatial and spectral resolution images of the surface mineralogy, provide information on the morphology of the Martian surface, and characterize the Martian near-space radiation environment.[23]

NASA's twin robot geologists, the Mars exploration rovers (MER), namely Spirit and Opportunity, were "launched toward Mars on June 10 and July 7, 2003, respectively, in search of answers about the history of water on Mars." One of the primary scientific objectives of the MER mission was "to search for and characterize a wide range of rocks and soils that hold clues to past water activity on Mars." The MER mission is part of NASA's Mars exploration program.

These small rovers with their various revelations of the landscapes of Mars, their miraculous recoveries and restarts, and other accomplishments have captured public attention and stirred interest in Mars in remarkable ways.[24]

In August 2005, NASA launched the Mars Reconnaissance Orbiter mission "on a search for evidence that water persisted on the surface of Mars for a long period of time."[25]

NASA's Phoenix mission was launched in August 2007 as part of the Mars exploration program with two objectives: "(1) study the history of water in the Martian arctic and (2) search for evidence of a habitable zone and assess the biological potential of the ice-soil boundary." The Phoenix Mars Lander ended operation in May 2010.[26]

NASA launched the Mars Atmosphere and Volatile EvolutioN (MAVEN) mission in November 2013 to explore Mars's "upper atmosphere, ionosphere and interactions with the sun and solar wind."[27]

The Mars Science Laboratory spacecraft, with the Mars rover Curiosity, was launched on November 26, 2011, and landed successfully on 6 August 6, 2012. The mission of Curiosity is to investigate "whether conditions have been favorable for microbial life and for preserving clues in the rocks about possible past life."[28]

NASA most current Mars mission is called InSight. This acronym stands for "Interior exploration using Seismic Investigations, Geodesy and Heat Transport." (see Fig. 5.2) This mission is a part of NASA's 2016 Discovery program. The venture intends to place a single geophysical lander on Mars to study its deep interior." InSight is "a terrestrial planet explorer that will open a window into the processes that shaped the rocky planets of the inner Solar System (including Earth) more than 4 billion years ago." The mission has two science goals and objectives:

[23] Objectives, 2001 Mars Odyssey, NASA http://mars.jpl.nasa.gov/odyssey/mission/science/objectives.

[24] Summary, Mars Exploration Rovers, NASA http://mars.nasa.gov/mer/overview.

[25] Mission Overview, MRO, NASA http://www.nasa.gov/mission_pages/MRO/mission/index.html#.VO5ZcS6sjEY.

[26] Phoenix Mars Mission, University of Arizona http://phoenix.lpl.arizona.edu/index.php.

[27] MAVEN: Answers About Mars' Climate History, MAVEN, NASA http://www.nasa.gov/mission_pages/maven/overview/index.html#.VO5p9i6sjEY.

[28] The rover's primary mission is to find out if Mars is, or was, suitable for life. Another objective is to learn more about the red planet's environment." Elizabeth Howell, "Mars Curiosity: Facts and Information", Space.com (16 December 2014), online: Space.com http://www.space.com/17963-mars-curiosity.html.

Fig. 5.2 InSight, the mars lander, shown in final assembly (Image courtesy of NASA.)

understanding "the formation and evolution of terrestrial planets through investigation of the interior structure and processes of Mars" and determining "the present level of tectonic activity and meteorite impact rate on Mars." The primary mission will end in September 2018.[29]

NASA has proposed a Mars 2020 mission, which is a part of NASA's Mars exploration program, that "would build upon many discoveries from the Curiosity Mars rover and the two Mars exploration rovers, Spirit and Opportunity, by taking the next key steps in our understanding of Mars' potential as a habitat for past or present life." The new Mars 2020 rover "would be designed to seek signs of past life on Mars, collect and store a set of soil and rock samples that could be returned to Earth in the future, and test new technology to benefit future robotic and human exploration of Mars." These functions could also support future efforts that might involve plans to create a habitat or to undertake mining operations.[30]

[29] NASA Facts: InSight…into the Early Evolution of Terrestrial Planets, JPL 400-1513, Rev 2 6/13 (Pasadena, Cal: Jet Propulsion Laboratory, California Institute of Technology) at 1, online: NASA http://insight.jpl.nasa.gov//docs/InSight_NASA_fact_sheet_rev3_June_2013_FC.pdf.

[30] NASA, NASA Facts: Mission Concept: Mars 2020 (October 2013) at 1, online: NASA http://mars.jpl.nasa.gov/mars2020/files/mars2020/FINAL_Mars_2020_handout_10-7-13.pdf.

NASA Asteroid Mission

The first mission to orbit an asteroid as well as the first to touch down on the surface of an asteroid is known as NASA's Near Earth Asteroid Rendezvous—the Shoemaker (NEAR) mission. The NEAR mission was launched on February 1996 to study the NEA 433 Eros. The primary scientific objectives of the mission were "to return data on the bulk properties, composition, mineralogy, morphology, internal mass distribution and magnetic field of Eros." The mission ended on February 2001.[31]

NASA also has plans to launch the Origins Spectral Interpretation Resource Identification Security—Regolith Explorer (OSIRIS-REx) mission in late 2016. The OSIRIS-REx spacecraft will "travel to a near-Earth asteroid, called Bennu (formerly 1999 RQ36), and bring at least a 2.1-ounce sample back to Earth for study."[32]

NASA launched the Stardust mission in February 1999. This was the first U. S. mission "dedicated solely to a comet." Stardust was directed toward comet Wild-2 with the principal objective to capture a sample from that comet. The Stardust mission was extended to fly by comet Tempel 1 in February 2011 with a new mission title, namely Stardust-NExT (Stardust-New Exploration of Tempel), following the sample return from comet Wild-2.[33]

NASA launched the combined Deep Impact spacecraft in January 2005 that was also directed toward comet Tempel 1.[34] The Deep Impact mission was "the first space mission to probe beneath the surface of a comet and reveal the secrets of its interior." The supplemental mission to explore other celestial targets by the Deep Impact spacecraft was dubbed EPOXI. On November 1, 2007, the spacecraft was directed toward another comet, Hartley 2. The spacecraft was also used as a test platform for a delay-tolerant networking transmission while at a distance of 20 million miles from Earth. Experimentation with long latency transmissions is one of the key technologies that will eventually be crucial for mining of asteroids.

The mission ended on September 19, 2013, due to a computer failure that doomed the spacecraft. At one point it had been intended for this spacecraft to fire its thrusters to target (163249) 2002 GT, an NEA, in hopes of intercepting it for study in 2020. Stardust-NExT mission extended the investigation of comet Tempel 1 by the Deep Impact mission, but the NEA intercept was not accomplished.[35]

The Submillimeter Wave Astronomy Satellite (SWAS) mission was launched in December 1998 by NASA with the primary objective "to survey water, molecular oxygen, carbon, and isotopic carbon monoxide emission in a variety of galactic star forming regions." The

[31] NEAR-Shoemaker, NASA Science, NASA http://science.nasa.gov/missions/near.

[32] OSIRIS-REx, NASA http://www.nasa.gov/mission_pages/osiris-rex/index.html#.VO-Ujy6sjEY.

[33] Mission Overview, Stardust-NExT, NASA http://www.nasa.gov/mission_pages/stardust/mission/index.html.

[34] Deep Impact: Mission to a Comet, NASA http://www.nasa.gov/mission_pages/deepimpact/mission/index.html#.VN_aHfnF-So.

[35] Deep Impact: Mission to a Comet, NASA http://www.nasa.gov/mission_pages/deepimpact/mission/index.html#.VN_aHfnF-So.

SWAS mission supported the Deep Impact mission.[36]

Assessing the Broad Impact of U. S. Space Missions Over the Past Half Century

The U. S. space missions over the past half century have contributed a great deal of understanding concerning the nature of the universe. Some of these have been in partnership with the Department of Defense or other agencies such as the National Science Foundation. These activities has revealed a wealth of information on the Moon, all of the planets and some of the moons, asteroids and comets as well as more remote star systems and exoplanets. Along the way U. S. space activities have aided in the development of space launcher and thruster systems and spaceplanes, satellite communications, remote sensing, space navigation and stabilization and guidance systems. U. S. space missions have revealed important findings on space radiation, coronal mass ejections, the magnetic characteristics of planets and moons, and the chemistry, soil composition, internal physics and biochemistry of the Solar System. All of the technologies and research findings are widely available around the world, on websites, to industry, which can indeed provide useful pathways to the future for many new space ventures. In short, a diverse range of U. S. governmental space activities—especially those of NASA—has created a vital knowledge foundation for new enterprises such Planetary Resources, Deep Space Industries, Shackleton Energy Company, Moon Express and other similar ventures that may spring up around the world in the form of either new commercial space enterprises or governmental space activities and missions. As will be seen in following chapters, a tremendous amount of information has also been contributed by the space programs of many spacefaring nations, but the U. S. space program has provided unparalleled data and scientific findings.

The following section seeks to assess the significance of some of these activities.

Space Telescope Findings

The Hubble Space Telescope was the first instrument to glimpse the atmosphere of a planet outside the Solar System. It also confirmed that the atmosphere of a particular exoplanet contained sodium, hydrogen, oxygen, carbon, silicon, water vapor, methane and carbon dioxide. Hubble "made the first detection ever of an organic molecule in the atmosphere of a Jupiter-sized" exoplanet. It has also confirmed the present of water vapor in four other exoplanets, namely WASP-17b, WASP-12b, WASP-19b and XO-1b. In the atmosphere of extrasolar planet WASP-12b, Hubble identified a wide variety of ingredients: aluminum, tin, magnesium, sodium, manganese, ytterbium, scandium and vanadium.[37] Within the Solar System it has allowed us to make a wide

[36] SWAS, Harvard University https://www.cfa.harvard.edu/swas/swas.html.

[37] Discovering Planets Beyond, Hubble Discoveries, Hubblesite http://hubblesite.org/hubble_discoveries/discovering_planets_beyond/alien-atmospheres.

ranging number of discoveries such as four new moons of Pluto, and with the deployment of the James Webb Space Telescope in the near future we could observe objects such as NEAs with a precision and resolution never before possible.

The Kepler spacecraft has discovered on the order of 10,000 actual confirmed planets, candidate planets or eclipsing binary stars that could yield planets, and opened the door to a finding a record number of exoplanets.[38]

Useful Information about the Moon

Data derived from Clementine mission enabled "the global mapping of the rock types of the lunar crust and the first detailed investigation of the geology of the lunar polar regions and the lunar far side." It also provided significant evidence of the existence of frozen water in the Moon's craters. Analyses of neutron data from the LP mission demonstrated the existence of hydrogen, at both poles of the Moon and that "at least some of the enhanced hydrogen deposits … are most likely in the form of water molecules."[39]

The LRO and LCROSS missions discovered evidence for water and hydrogen in certain places on the Moon. The two missions "found evidence that the lunar soil within shadowy craters is rich in useful materials, and the Moon is chemically active and has a water cycle." These missions also confirmed that the water was in the form of mostly pure ice crystals in some places. It was determined that "as much as 20 % of the material kicked up by the LCROSS impact represented volatiles, including methane, ammonia, hydrogen gas, carbon dioxide and carbon monoxide." These missions also confirmed that large amounts of metals such as sodium, mercury and possibly even silver were also present. [40]

The LRO and LCROSS missions led to the conclusion that "Lunar geology can be roughly broken down into two categories—the anorthositic highlands," which are "rich in calcium and aluminum and the basaltic "maria." These maria, or giant impact basins, are filled with solidified lava flows, which are plentiful in iron and magnesium. The LRO mission further confirmed that "most lunar terrains have signatures consistent with compositions in these two broad categories. Highly silicic minerals such as quartz, potassium-rich and sodium-rich feldspar, were detected in several locations around the Moon.[41]

Finally NASA's LADEE mission confirmed that there is a more or less steady rain of micrometeoroid particles onto the lunar surface," and that "the major gas species in the thin lunar atmosphere are … helium, neon and argon," particularly argon-40, which "comes from the decay of naturally occurring radioactive potassium-40, found in the rocks of all the terrestrial planets as a leftover from formation." Two minor species, namely sodium and potassium, were also monitored. Further the

[38] Kepler Discoveries, Kepler, NASA http://kepler.nasa.gov/Mission/discoveries.

[39] The Clementine Mission, http://www.lpi.usra.edu/lunar/missions/clementine/.

[40] Ice on the Moon http://nssdc.gsfc.nasa.gov/planetary/ice/ice_moon.html.

[41] *Ibid.*

presence of the rare and valuable helium-3 was confirmed.[42]

Findings Related to Planetary Bodies

The *Pioneer 10* mission "established that Jupiter is predominantly a liquid planet." It also measured Jupiter's "magnetosphere, radiation belts, magnetic field, atmosphere, and interior." The Voyager mission discovered active volcanoes on one of the Jupiter's moon, namely Io. Apparently, sulfur, oxygen and sodium were emitted by those volcanoes. Another moon, namely Europa, is "thought to have a thin crust … of water ice." Notable discoveries related to natural resources of NASA's Galileo mission include some evidence for a liquid water ocean under Jupiter's moon Europa's icy surface, and organic compounds on Jupiter's moons Callisto, Ganymede, Europa and Io.[43]

With respect to Saturn, the Voyager mission discovered that, in Saturn's moon Titan, photochemistry "converts some atmospheric methane to other organic molecules, such as ethane, that is thought to accumulate in lakes or oceans." It was also found that, in Titan, it was likely that "more complex hydrocarbons form the haze particles." The Cassini-Huygens mission to Saturn discovered a significant number of facts about Saturn and its two moons, Titan and Enceladus. According to the Cassini spacecraft, Enceladus has an underground sea of liquid water. Previously, the Cassini-Huygens mission "directly sampled the water plumes jetting into space from … Enceladus," which was the strongest evidence yet for the existence of large-scale saltwater reservoirs beneath Enceladus's icy crust.[44] Perhaps most spectacularly, the Cassini spacecraft sensors observed "what appears to be a miniature extraterrestrial version of the Nile River." This seemed to be a river valley on Saturn's moon Titan that stretched more than 400 km from its 'headwaters' to a large sea." Earlier, evidence for seas likely filled with liquid methane and ethane, lakes filled with liquid methane, and large tides were found on Titan. The large tide findings pointed to "a liquid ocean—most likely water—swirling around below the surface." It was also found that Titan has "hundreds of times more liquid hydrocarbons than all the known oil and natural gas reserves on Earth."[45]

Mariner 2, the first U. S. mission to Venus, revealed that this planet "has cool clouds and an extremely hot surface." *Mariner 5* revealed "new information about Venus' atmosphere,

[42] LADEE Project Scientist Update: The Legacy Lives On! https://www.nasa.gov/ames/ladee-project-scientist-update-the-legacy-lives-on/.

[43] Pioneer-10 and Pioneer-11" (26 March 2007), online: Mission Archives, NASA http://www.nasa.gov/centers/ames/missions/archive/pioneer10-11.html Also see Pioneer, Program, Jet Propulsion Laboratory http://space.jpl.nasa.gov/msl/Programs/pioneer.html. Also see:Galileo, Solar System Exploration, NASA http://solarsystem.nasa.gov/missions/profile.cfm?Sort=Chron&MCode=Galileo&StartYear=1980&EndYear=1989&Display=ReadMore.

[44] ESA, News, "Cassini Samples the Icy Spray of Enceladus' Water Plumes" (22 June 2011), ESA http://www.esa.int/Our_Activities/Space_Science/Cassini-.

[45] ESA, News, "Cassini Spots Mini Nile River on Saturn Moon" (12 December 2012), online: ESA http://www.esa.int/Our_Activities/Space_Science/Cassini_spots_mini_Nile_River_on_Saturn_moon.

including its composition of 85–99 % carbon dioxide." *Mariner 10*, which was the first spacecraft to visit two planets and the first to visit Mercury, collected important scientific data on Venus. The results of *Mariner 10* demonstrated, *inter alia*, that "a Hadley-type circulation existed in Venus' atmosphere." The *Pioneer 12* spacecraft returned global maps of clouds, atmosphere and the ionosphere of Venus, "measurements of the atmosphere-solar wind interaction, and radar maps of 93 % of the planet's surface." The *Pioneer 13* mission studied, *inter alia*, the atmospheric composition, composition of the upper atmosphere, and cloud particles of Venus. Among others, one of the key scientific findings of the Magellan mission was that planet Venus's surface is "mostly covered by volcanic materials. Volcanic surface features, such as vast lava plains, fields of small lava domes, and large shield volcanoes are common."

The *Mariner 10* mission "confirmed that Mercury had no atmosphere and a cratered, dormant Moon-like surface was shown in the images. Mercury was shown to have … a relatively large iron-rich core." The ongoing MESSENGER mission found evidence for "water ice and other frozen volatile materials in [Mercury's] permanently shadowed polar craters."[46]

Images from the *Voyager 2* spacecraft showed active geyser-like eruptions on Neptune's largest moon Triton, "spewing invisible nitrogen gas and dark dust particles several miles (km) into the tenuous atmosphere." In Triton nitrogen ice particles may form thin clouds a few km above the surface."

Mars Exploratory Programs

Mariner 4, which was the first U. S. spacecraft to get a close look at Mars, revealed that it has "a cratered, rust-colored surface, with signs on some parts of the planet that liquid water had once etched its way into the soil." *Mariners 6* and *7* revealed "cratered deserts, as well as depressions with no craters, huge concentrically terraced impact regions, and collapsed ridges." The images from *Mariner 9* revealed "river beds, craters, massive extinct volcanoes, canyons." It also found "evidence of wind and water erosion and deposition, weather fronts, fogs, and more." These findings "laid the groundwork for the Viking program."[47]

NASA's Viking mission "provided numerous new insights into the nature and history of Mars." It produced "a vivid overall picture of a cold weathered surface with reddish volcanic soil under a thin, dry carbon dioxide atmosphere, clear evidence for the existence of ancient river beds and vast floods, and no detectable seismic activity." The mission did not find any traces of life on Mars but found carbon, nitrogen, hydrogen, oxygen and phosphorus, which are essential to life on Earth.[48]

NASA's Mars Global Surveyor mission "contributed a multitude of

[46] Mariner 10, Solar System Exploration, NASA http://solarsystem.nasa.gov/missions/profile.cfm?MCode=Mariner_10&Display=ReadMore.

[47] Mariner Missions, NASA http://science1.nasa.gov/missions/mariner-missions.

[48] Viking 01, Solar System Exploration, NASA http://solarsystem.nasa.gov/missions/profile.cfm?Sort=Alpha&Letter=V&Alias=Viking%2001.

findings, including signs of past, persistent water such as an ancient delta and currently active water features in the gullies of canyon walls."[49]

Findings from the Mars Pathfinder mission "suggest that Mars was at one time in its past warm and wet, with water existing in its liquid state and a thicker atmosphere."[50]

The 2001 Mars Odyssey mission "mapped the amount and distribution of chemical elements and minerals that make up the [Martian] surface. Maps of hydrogen distribution led scientists to discover vast amounts of water ice in the Polar Regions just beneath the surface."[51]

There were many new discoveries from the Mars exploration twin rovers—Spirit and Opportunity. Some of these results include finding the mineral hematite that typically forms in water, finding "rocks ten times richer in key chemicals (magnesium and iron carbonates) than any other Martian rocks studied before," finding 90% pure silica, which, on Earth, "exists in hot springs or hot steam vents," finding "bright-colored veins of gypsum in the rocks … [which] likely formed when water flowed through underground fractures in the rocks, leaving calcium behind," and finding "the most compelling signs of a watery past on Mars in the form clay minerals formed in neutral-pH water."[52]

Significant findings of the Mars Reconnaissance Orbiter mission include signs of ancient Martian lakes and quakes, clues to possible water flows, evidence for carbonate forming within a lake environment, and evidence of carbon dioxide snowfalls on Mars.[53]

The Phoenix Mars lander uncovered a shallow ice table, and it was discovered that H_2O ice and vapor relentlessly interacted with the soil, which was alkaline, containing $CaCO_3$, aqueous minerals, and salts. Calcium carbonate, perchlorate (ClO_4), chloride, bicarbonate and sulfate were also detected.[54]

The Mars rover Curiosity "met its major objective of finding evidence of a past environment well suited to supporting microbial life." "In the first few weeks after landing, images from the rover showed that Curiosity touched down right in an area where water once coursed vigorously over the surface (Fig. 5.3)." During the rover's "approach to the base of Mount Sharp, the rover found sedimentary rocks fitting [the] pattern of successive delta deposits, with the ones farther south (toward the mountain) higher in elevation." This finding "suggests that where the mountain stands now was formerly a lake or

[49] Overview, Mars Global Surveyor, NASA http://mars.jpl.nasa.gov/mgs/overview/. To learn more about the top discoveries of Mars Global Surveyor mission, also see NASA, "NASA—Mars Global Surveyor Mission Highlights" (13 April 2007), Mars Global Surveyor, NASA http://www.nasa.gov/mission_pages/mgs/mgs-20070413a.html#.VO0VgS6sjEY.

[50] Overview, Mars Pathfinder, NASA http://www.nasa.gov/mission_pages/mars-pathfinder/index.html.

[51] Science, 2001 Mars Odyssey, NASA http://mars.jpl.nasa.gov/odyssey/mission/science/.

[52] Science Highlights, Mars Exploration, NASA http://mars.jpl.nasa.gov/mer10/sciencehighlights.

[53] "NASA Orbiter Observations Point to 'Dry Ice' Snowfall on Mars" (11 September 2012), online: Mars Reconnaissance Orbiter, NASA http://www.nasa.gov/mission_pages/MRO/news/mro20120911.html.

[54] MH Hecht et al, "Detection of Perchlorate and the Soluble Chemistry of Martian Soil at the Phoenix Lander Site" (2009) 325:5936 Science 64.

Fig. 5.3 The mars curiosity rover (Image courtesy of NASA.)

series of lakes." The rover found evidence of an ancient streambed. The first drilled sample from a rock target called "John Klein" was analyzed, and it "provided the evidence of conditions favorable for life in Mars' early history: geological and mineralogical evidence for sustained liquid water, other key elemental ingredients for life, a chemical energy source, and water not too acidic or too salty." Later, analysis of a sample from another rock target called "Cumberland" "yielded the first definitive detection of any Martian organic chemicals in material on the surface of Mars." Organic chemicals "are molecular building blocks of life," containing carbon and usually hydrogen.[55]

[55] Mike Wall, "Wow! Ancient Mars Could Have Supported Primitive Life, NASA Says", Space.com (12 March 2013), Also see: Space.com http://www.space.com/20182-ancient-mars-microbes-curiosity-rover.htm.

Missions to Comets and Asteroids

The NEAR-Shoemaker mission enabled scientists to determine that the asteroid Eros "is not a "rubble pile" of loosely bound pieces, but rather a consolidated object."[56]

The Stardust spacecraft was "the first to return extraterrestrial material from outside the orbit of the Moon." Glycine, a fundamental building block of life, was discovered in samples of comet Wild 2. High-temperature minerals rich in magnesium, calcium, aluminum and titanium were also found.[57]

[56] NEAR-Shoemaker, NASA Science, NASA http://science.nasa.gov/missions/near/.

[57] NASA, News Release, 06-091, "NASA's Stardust Findings May Alter View of Comet Formation" (13 March 2006), online: Newsroom, NASA http://stardust.jpl.nasa.gov/news/status/060313.html.

With respect to natural resources, one of the key findings of the Deep Impact mission, which includes the EPOXI mission, is that "comet tails contain dry ice and water."

The recent NASA Dawn mission to explore Ceres and Vesta was designed to understand better the formation, composition, and nature of asteroids. During its nearly decade-long mission, Dawn will study in some detail both the asteroid Vesta and dwarf planet Ceres. These asteroid bodies are believed to have accreted early in the history of the Solar System. The mission will characterize the early Solar System and the processes that led to its formation.

The current theory is that in the earliest epochs of our Solar System, the materials in the solar nebula varied considerably as a function of their distance from the Sun. As this distance increased, the temperature dropped, with terrestrial bodies forming closer to the Sun, and icy bodies forming farther away.

The asteroid Vesta and the recently categorized dwarf planet Ceres were selected because they developed into two different kinds of bodies even though similar distances from the Sun. Vesta is a dry, differentiated object with a surface that shows signs of resurfacing. It resembles the rocky bodies of the inner Solar System, including Earth. Ceres, by contrast, has a primitive surface containing water-bearing minerals, and may possess a weak atmosphere. It appears to have many similarities to the large icy moons of the outer Solar System. The results of the Dawn mission should help explain these significant differences and why they may have occurred.[58]

[58] Dawn Mission Overview http://www.nasa.gov/mission_pages/dawn/mission/index.html.

The Future of Space Exploration Technology Related to Space Mining

The largest future contribution that might be made by the U. S. space program to future space mining activities would be more and better information about NEAs in terms of their detection and orbital mapping as well as more information about the chemical and physical attributes of these objects. Growing concerns about the cosmic hazards constituted by PHAs could accelerate efforts in this respect. The Sentinel project by the B612 Foundation, the NASA proposed NEOCAM project, the expanded capabilities represented by the James Webb Space Telescope as well as other capabilities such as Canadian infrared telescope for finding PHAs could all support future space activities aimed at mining asteroids.

Conclusions

Clearly the U. S. governmental space program as led by NASA for over 50 years has created a large and systematic base of information about the chemical and physical composition of the Moon, asteroids, comets and planetary bodies. This information substantiates that there is out in space a wealth of volatiles, precious and rare earth metals that are becoming more and more difficult to locate and economically mine here on Planet Earth. The wealth of information that has been amassed in hundreds of different NASA and other U. S. agencies over a half century has now established a base from which new entrepreneurial efforts targeted toward space mining can effectively draw. NASA efforts via Space Act Agreements, mechanisms such as the Lunar Catalyst, design and engineering contests related to robotic mining, tele-robotics, remote and high latency communications and networking systems, and precision space navigation and guidance systems can also help provide critical new information.

A number of legal, regulatory and policy issues have yet to be addressed and resolved before such activities can commence. As these issues are resolved in the next decade, the scientific, engineering, research and technical efforts to develop the needed capabilities will continue to move forward with support from the U. S. space program and that of other space agencies such as Roscosmos in Russia, ESA, CNES, DLV and so on in Europe, JAXA in Japan, the Chinese National Space Agency, the Indian Space Research Organization, the Canadian Space Agency and other space programs in other parts of the world.

6 Private Sector Space Mining Initiatives and Policies in the United States

The development of a dynamic commercial so-called "new space" industry in the United States over the past 15 years has redefined the respective roles of private versus publicly funded space programs. The initiatives of private entities have created new perspectives on what types of space services can be offered by commercial space enterprises and opened eyes to fresh new technilogical and operational approaches to space services and systems. In addition to a wide range of new commercial space ventures new organizational arrangements have emerged. Prime among these new institutions is the Commercial Spaceflight Federation (CSF) that has a wide membership among commercial spaceflight companies, spaceports, and others seeking to develop new space activities.

These new space initiatives have also served to change U. S. regulators views on how to control, license and encourage commercial space innovation. Various "space acts" passed by Congress and signed into law have by and large served to encourage new commercial space enterprise. Today the U. S. Federal Aviation Administration Office of Commercial Space Transportation (FAA-AST) serves a dual role of regulator of commercial space safety as well as encourager of new space enterprises.

Today NASA also relies on commercial vehicles to access the International Space Station and have awarded contracts for new vehicles and capsules to take astronauts into and return from space. Commercial ventures have developed spaceplanes to provide suborbital tourist experiences. Commercial vehicles are now being developed to place satellites and even people into space as well as to deploy private space habitats. Robotic systems are being developed to refuel and service spacecraft and even actively deorbit space debris. Some of these systems could even be deployed as anti-satellite weapons. Also, the U. S. Department of Defense and especially its Defense Advanced Research Projects Agency (DARPA) plays a key role in developing new capabilities in space, frequently in partnership with private space ventures.

This dramatic shift in the division between private and public space programs and the rapid rise of private space systems to carry out activities that are sometimes called new space activities

© Springer International Publishing Switzerland 2017

R.S. Jakhu et al., *Space Mining and Its Regulation*,
Astronautical Engineering, DOI 10.1007/978-3-319-39246-2_6

has been driven first and foremost in the United States. This chapter explores the many new space programs that have been started—largely by entrepreneurial and startup aerospace companies in the United States in the last 15 years. It then proceeds to examine the various new space enterprises that are being formed to pursue space mining and the recovery of natural resources from space. Finally this analysis covers the planned course of action plus the ambitious goals and objectives of new U. S. space ventures. It covers their efforts to create a flexible partnership with U. S. governmental space agencies as well as to forge an open and permissive regulatory structure for their operations. Their plans seek for them to be able to proceed with a minimum of regulatory oversight, with few restrictions other than for the safety of their operations and with very liberal interpretation of existing space law, treaties and conventions.

The Rapid Growth of New Space Activities in the United States

Several U. S. entrepreneurs are developing new space ventures with the goal of carrying out space mining as soon as within another decade. These American space entrepreneurs strongly believe that the private sector will play a major role in new space initiatives, and they are intent on being the pioneers that make this actually happen. They are advocating private enterprise take the lead and are correspondingly advocating a limited role for government. A number of these individuals and their companies have in the past strongly supported active involvement of the private sector in other space activities. These space and "protospace" ventures have included such activities as zero g flights, suborbital "space tourism" flights, stratospheric balloon flights, high altitude platform systems for various applications, and private astronaut flights to space habitats and the International Space Station. Their latest initiatives to pursue private ventures aimed at the exploitation of space's natural resources is very much a logical thought extension of their earlier efforts to "privatize" space activities.

In the short to medium term, those private actors may not be able to mount a completely comprehensive effort to provide all the needed space exploration and exploitation technologies to cover all aspects of a space mining enterprise. They have therefore sought to get the U. S. government to help them develop new capabilities. They have first worked to create new mechanisms such as the Commercial Spaceflight Federation (CSF) to strengthen their voice. They have encouraged NASA and the FAA-AST to sponsor contests and put up prize money to stimulate new commercial space competency. Their aspiration is to do as much of these new space activities—such as space mining and space transportation—as soon as possible via commercial mechanisms and with the minimum amount of governmental involvement and regulation.

Despite these aspirations there is some recognition on their part that cooperative relationships with governmental space agencies, research agencies, and national and international regulatory bodies may still be necessary. They understand that the substantial costs, the need for certain technilogical capabilities, risk management, and international regulatory controls may ultimately require governmental cooperation at the national level as well as in the international space governance arena.

However, in the United States—and the phenomenon is currently largely limited to this country—the role and voice of these new space entrepreneurs and their various space mining ventures are increasing heard within Congress and the U. S. federal government—at least when it comes to governmental space policy. These individuals and associated new space businesses often have access to substantial amounts of financial capital and in various ways have had a remarkable impact on recent space policies adopted by the government. In short they have impacted the mind set of congressional legislators and staff as well as various Executive Branch and even local state officials.

Just a few of the most recent privately led initiatives are the X-Prize (to build a private, reusable spaceship that will herald a new era in commercial spaceflight); [1] the Google Lunar X Prize (to send a robot to the Moon and perform a series of tasks; the SpaceX project (to develop the first private sector launcher)[2]; and, the Bigelow Aerospace project (which has deployed an inflatable module in low Earth orbit that could become the space habitat of the future).[3]

And NASA and the FAA have been increasingly supportive of new private space initiatives. NASA, for instance, created a venture with RedPlanet Capital (with an investment of $75 million) at the end of 2006 to develop technology that could help the agency to send missions to Mars. The aim was to find companies whose technologies could also represent significant breakthroughs on Earth, as well as in the heavens.[4] NASA has also embarked on a series of prize competitions to develop new technologies for Moon and Mars landings and even to develop the capabilities to design and build space elevators.[5] NASA has sought in the past decade to develop new commercial rockets to resupply the International Space Station. The first step was to develop commercial resupply vehicles. The first commercial competition started in 2006 with awards to SpaceX and Kistler Aerospace. When Kistler was not able to meet performance deadlines NASA shifted the

[1] The X-Prize was modeled on the Orteig Prize offered in 1919 to the first pilot who could cross the Atlantic in a non-stop flight. Charles Lindbergh won the price in 1927. See The X-Prize Foundation online: http://www.xprize.org/ (Date accessed: 13 March 2013). "Virgin Galactic unveils SpaceShipTwo, the world's first manned commercial spaceship" (December 7, 2009) online: Virgin Galactic: http://www.virgingalactic.com/news/item/virgin-galactic-unveils-spaceshiptwo-the-worlds-first-commercial-manned-spaceship/ (Date accessed: March 13, 2013).

[2] For an overview of the SpaceX company, see online: SpaceX http://www.spacex.com/company.php (Date accessed: March 13, 2013).

[3] George Knapp, "The ultimate public-private partnership" The Las Vegas Mercury (July 8 2004) online: The Las Vegas Mercury http://www.lasvegasmercury.com/2004/MERC-Jul-08-Thu-2004/24250261.html (Date accessed: March 13, 2013).

[4] Stephen Foley, "NASA seeks private investor backing for mission to Mars" The Independent [UK] (October 4 2006) online: The Independent http://www.independent.co.uk/news/business/news/nasa-seeks-private-investor-backing-for-mission-to-mars-418648.html (Date accessed: March 13, 2013).

[5] Graham Templeton, "60,000 miles up: Space Elevator could be built by 2035" Extreme Tech, March 6, 2014: http://www.extremetech.com/extreme/176625-60000-miles-up-geostationary-space-elevator-could-be-built-by-2035-says-new-study (Date accessed: August 16, 2015).

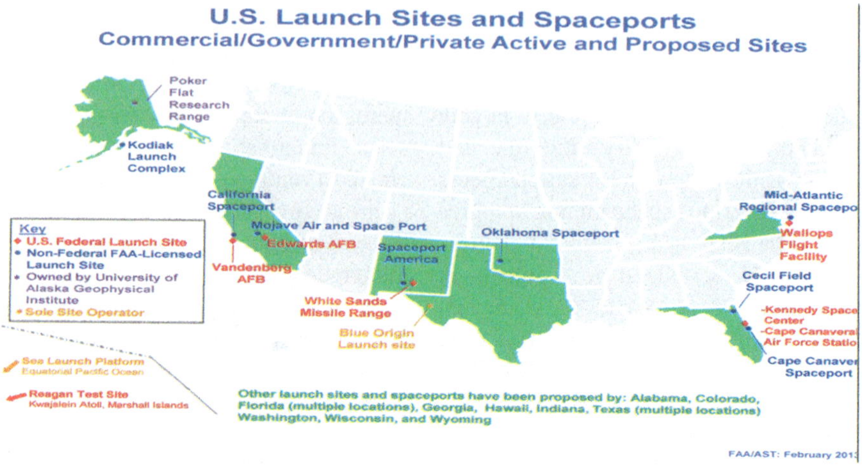

Fig. 6.1 FAA-licensed and pending commercial spaceports in the United States (Graphic courtesy of the U. S. Federal Aviation Administration.)

award to Orbital Sciences (now Orbital ATK). This effort to develop a commercial orbital transportation service evolved into a NASA program to create a commercial capsule and launchers that are being developed by SpaceX and Boeing under two multi-billion-dollar contracts.[6]

The FAA Office of Commercial Space Transport has created under congressional guidance regulatory processes for granting experimental licenses for commercial suborbital flights that has been in many ways quite flexible. Even more significant is that FAA-AST has licensed a growing number of commercial spaceports in the United States. The number of commercial spaceports that have been licensed in the United States (plus those currently pending license approval) far outnumber the commercial spaceports in the rest of the world by a wide margin. Figure 6.1 shows in blue dots the commercial spaceports fully licensed plus indicates the states where about a dozen spaceports are pending license approvals.[7]

The number of commercial launches that have been approved under experimental licenses by the FAA is now quite significant and also rapidly increasing. Thus not only are there many more commercial spaceports, but there have been far more commercial launches in the United States than anywhere else in the world. Table 6.1 below provides a listing of experimental licenses given to commercial launch developers between 2008 and 2015. Beyond this list of actual launches under experimental licenses there are more than a dozen other U. S.

[6]NASA, Boeing, SpaceX Discuss Plan for Launching American Astronauts from U.S. in 2017: http://www.nasa.gov/centers/kennedy/news/release-20150121.html (Date accessed: August 17, 2015).

[7]FAA chart of licensed US commercial spaceports https://www.faa.gov/about/office_org/headquarters_offices/ast/industry/media/Spaceport_Map_Feb_2013.pdf.

Table 6.1 Experimental U. S. launches during the 2008–2015 period (Information courtesy of the FAA.)

Date	Payload	Vehicle	Company	Site
4-29-2015	None	New Shepard System	Blue Origin	Texas
10-31-2014	None	SpaceShipTwo	Scaled Composites	CA
8-22-2014	N/A	Falcon 9-R	Space X	Texas
8-21-2014	None	Falcon 9-R	Space X	Texas
6-17-2014	None	Falcon 9-R	Space X	Texas
5-1-2014	N/A	Falcon 9-R	Space X	Texas
4-17-2014	N/A	Falcon 9-R	Space X	Texas
1-10-2014	Flight PF03	SpaceShipTwo	Scaled Composites	CA
10-7-2013	750 M	Grasshopper	Space X	Texas
9-5-2013	Flight PF02	SpaceShipTwo	Scaled Composites	CA
8-13-2013	N/A	Grasshopper	Space X	Texas
6-14-2013	N/A	Grasshopper	Space X	Texas
4-29-2013	Flight PF01	SpaceShipTwo	Scaled Composites	CA
4-19-2013	N/A	Grasshopper	Space X	Texas
3-7-2013	N/A	Grasshopper	Space X	Texas
12-17-2012	N/A	Grasshopper	Space X	Texas
11-1-2012	No Payload	Grasshopper	Space X	Texas
8-24-2011	N/A	PM 2	Blue Origin	Texas
6-6-2011	N/A	PM 2	Blue Origin	Texas
10-25-2008	N/A	QUAD (Pixel)	Armadillo Aerospace	New Mexico
10-24-2008	N/A	MOD-1	Armadillo Aerospace	New Mexico
10-24-2008	N/A	MOD-1	Armadillo Aerospace	New Mexico
10-24-2008	N/A	Ignignokt	ScottZeeb d/b/a True ZerO	New Mexico

companies that are at various stages of developing commercial launchers or spaceplanes. Thus there may be a large increase in commercial launches under experimental licenses in the 2016–2018 timeframe.

The wide range of U. S. commercial space activities since 2000 has included development of high-altitude platform systems, stratospheric balloon systems, spaceplanes for suborbital flights, and commercial launchers capable of achieving low Earth orbit and beyond. These new space ventures are an extension of commercial space activities in telecommunications, remote sensing, and satellite navigation. Today there are emerging new commercial space industries such as on-orbit servicing, on-orbit refueling and retrofitting of satellites, and even commercial monitoring and active removal of space debris that are also predominately U. S.-based efforts as well. These latest efforts have been supported not only by NASA and the FAA-AST, but very prominently promoted by the Defense Advanced Research Projects Agency (DARPA). DARPA projects, such as Orbital

Express[8] and Project Phoenix[9], as well as joint development projects with NASA are heavily dependent on aerospace contractors.

These initiatives to develop private and more cost effective commercial launchers, to develop in-orbit robotic capabilities to refuel and service satellites, as well as other new commercial systems to maneuver remotely and precisely in space are all useful precursor technologies that could lead the way to future commercial space mining activities. Already several U. S.-based space mining activities and private asteroid tracking activities have been organized.

Planetary Resources

On April 24, 2012, Planetary Resources[10] became the first private enterprise to enter into the realm of private space business directly dealing with the quest for natural resources in of space. The vision of the company is as follows: "Planetary Resources is bringing the natural resources of space within humanity's economic sphere of influence, propelling our future into the twenty-first century and beyond. Water from asteroids will fuel the in-space economy, and rare metals will increase Earth's GDP."[11] The company wants to develop a low-cost robotic spacecraft to explore the estimated 9000 NEAs for potential resource extraction and utilization.[12] According to Peter Diamandis, the founder and co-chairman of Planetary Resources, "Many of the scarce metals and minerals on Earth are in near-infinite quantities in space. As access to these materials increases, not only will the cost of everything from microelectronics to energy storage be reduced, but new applications for these abundant elements will result in important and novel applications."[13] Eric Anderson, co-founder and co-chairman of the company, has also indicated that the first targets for exploration will be water-containing asteroids. "Water is perhaps the most valuable resource in space. Accessing a water-rich asteroid will greatly enable the large-scale exploration of the Solar System. In addition to supporting life, water will also be separated into oxygen and hydrogen for breathable air and rocket propellant."[14] The company has begun a detailed exercise to identify potential candidate

[8] DARPA Orbital Express Fact Sheet, http://archive.darpa.mil/orbitalexpress/pdf/oe_fact_sheet_final.pdf (Date accessed: August 19, 2015).

[9] Mike Wall, "US Military's Bold Phoenix Satellite-Recycling Project Enters New Phase" Space.com April 25, 2014.

[10] Planetary Resources, Inc. Press Conference, April 24, 2012. Link to the official announcement available from the company website, online: http://www.planetaryresources.com/.

[11] Planetary Resources, online: http://www.planetaryresources.com/mission/.

[12] Planetary Resources, "Asteroid mining plans revealed by Planetary Resources, Inc.: Expanding the resource base of humanity to include the solar system", Planetary Resources, April 24, 2012, online: http://www.planetaryresources.com/2012/04/asteroid-mining-plans-revealed-by-planetary-resources-inc/.

[13] Planetary Resources, "Asteroid mining plans revealed by Planetary Resources, Inc.: Expanding the resource base of humanity to include the solar system", Planetary Resources, 24 April 2012, online: http://www.planetaryresources.com/2012/04/asteroid-mining-plans-revealed-by-planetary-resources-inc/ (Last accessed: March 13, 2013).

[14] *Ibid.*

Fig. 6.2 Image of stony Eros 433—possible candidate for space mining? (Image courtesy of NASA.)

NEAs that could be reached without excessively high thrust impulse and would potentially contain natural resources of significant value. Its website includes a listing of dozens of potential target asteroids. It has also created a process where amateur astronomers and scientists can add information to their database.

To the uninitiated this might seem like a straightforward and not too complicated exercise, but in fact it is a major challenge. It is estimated that there may be a million NEAs that are 30 m or more in diameter. In Chap. 4 the many types of orbits that these NEAs travel in and the difficulty in locating and assessing their resource content was discussed.

However, this asteroid identification process is now actively underway by PRI. When the best "goldilocks" NEA candidate is identified—i.e., an asteroid that has the best resource content and in an orbit that would be not too difficult to access and mine, then Planetary Resources intends to launch its first asteroid-hunting spacecraft on a reconnaissance mission (Fig. 6.2). Currently the company is concentrating on the idea of creating small 3D-printed low-cost spacecraft to serve as the explorer-prospecting units that could fly close enough to an NEA to assess whether it is indeed a prime candidate for space mining.[15]

Deep Space Industries

On January 22, 2013, a second U. S.-based company, Deep Space Industries (DSI), entered the race of asteroid surveying and

[15] M. H., "Fool's platinum?" *The Economist*, 1 February 2013, online: http://www.economist.com/blogs/babbage/2013/01/asteroid-mining (Last accessed: March 13, 2013).

resource extraction.[16] The company intends to develop a fleet of three spacecraft using off-the-shelf technology to survey small NEAs.[17] It hopes to attract $13 million in capital over the next few years.[18] The Deep Space Industries website proclaims: Our mission is a daring one. We are journeying to unknown frontiers, and pushing the limits of technology to provide a brighter future for all mankind." Essentially, DSI is suggesting that there is great wealth in our Solar System and that their vision is to help bring that wealth back to Earth. They anticipate the following progression. After prospecting missions have identified asteroids with concentrated volatiles (such as water and hydrocarbons) and other materials of interest, Deep Space will begin collection with specialized robotic spacecraft. Deep Space thus describes its activities as a four-step progression that begins with prospecting, moves on to harvesting, then processing, and finally manufacturing. The aspirations of the company are far from small in scope. It has proclaimed that this will be the biggest industrial transformation in human history. (See https://deepspaceindustries.com/business/.)

To commence its activities, the company plans to send "asteroid-prospecting spacecraft" into outer space with the launch of the first of its 25-kg "FireFlies" spacecraft. This will be followed by the heavier "DragonFlies" that will go on mission and bring back samples. Deep Space Industries has trademarked the name Mothership™. This refers to the concept of a larger carrier spacecraft designed to aid in the delivery of nano satellites to deep space targets. After deploying the nanosats, the spacecraft remains as a high bandwidth communication relay between the deployed craft and Earth. One of the elements needing clarification is whether DSI sees its mission as mining asteroids or whether it also has aspirations to carry out mining on the Moon (Fig. 6.3).[19]

DSI has suggested that a spacecraft might capture an asteroid and reposition it in an orbit near Earth for potential harvesting of resources. In this concept there is significant reliance on solar power systems, but in other more conventional concepts, more proven chemical propulsion is envisioned.

In press statements DSI has indicated that even a small asteroid might ultimately be valued at up to $195 billion. This would represent, however, an extreme case of an NEA that is almost pure platinum. [20]

Golden Spike Company

On December 6, 2012, Golden Spike Company released its vision regarding the future of commercial lunar space travel.[21] It stated that it intends to achieve

[16] "US company aims to 'harvest' asteroids", 22 January 2013, online: http://www.spacedaily.com/reports/US_company_aims_to_harvest_asteroids_999.html (Last accessed: March 13, 2013).

[17] Deep Space Industries website: http://deepspaceindustries.com (Date accessed: August 2015).

[18] Ibid.

[19] http://deepspaceindustries.com/mothership/.

[20] "Deep Space Industry Asteroid Mining Plan," *Business Insider*: http://www.businessinsider.com/deep-space-industry-asteroid-mining-plan-2013-2 (Date accessed: August 20, 2015).

[21] Golden Spike Company Vision: "The Golden Spike Company has been formed to monetize the exploration of the Moon through sales of expeditions and their surrounding media and merchandizing revenues." See the company vision on the website: http://goldenspikecompany.com/about-us/golden-spike-history/ (Last accessed: March 13, 2013).

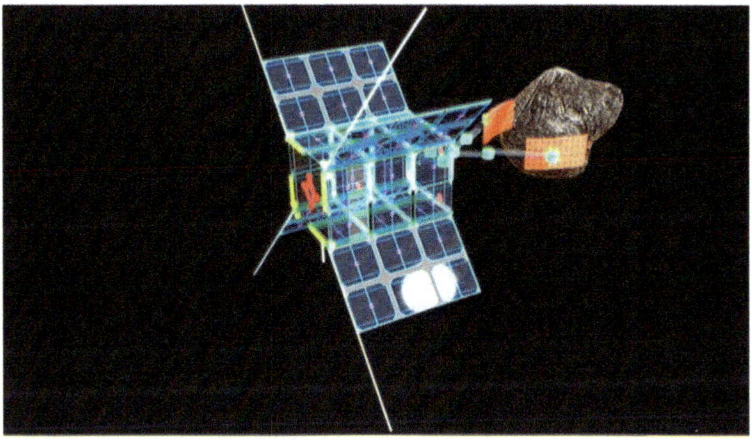

Fig. 6.3 Artist's conception from deep space industries of spacecraft carrying out an asteroid capture (Illustration courtesy of deep space industries.)

"affordable, reliable, and frequent human expeditions to the Moon."[22] Unlike other space tourism projects that involve short suborbital trips or possibly trips to low Earth orbit, this corporate venture, based in Colorado, envisions an actual trip to the Moon for two passengers at a cost of $750,000 per traveler. Its stated ambition is to be able to do this by 2020. This private space transportation capability could be used for prospecting the lunar surface or even as a prelude to setting up of a lunar colony. Although $1.5 billion per flight is a large amount of money it is a small expense in comparison to the Apollo lunar landing programs by the United States during the 1960s and early 1970s or even the $140 billion International Space Station project. It could be possible that a national or regional space agency could actually contract with such a private enterprise to initiate a new lunar exploration effort that might include the start of a lunar colony or even a lunar mining effort.

In January 2013, Golden Spike contracted with Northrop Grumman for the design of a new lunar lander, as one component of its initial commercial lunar transportation system. Contracted tasks include "reviewing requirements and synthesizing a set of study ground rules and assumptions emphasizing system reliability, automated/ground command operability, and affordability, establishing velocity (Δv) budgets from and to low lunar orbit for pragmatic lunar landing sites, exploring a wide variety of lunar lander concept options, including staging, propellants, engines, reusability, autonomy, systems capabilities for exploration, as well as landing site flexibility, and establishing the design trade space and pragmatic limits

[22] http://goldenspikecompany.com/our-business/business-objectives/ (Date accessed: April 2, 2013).

for future more detailed analysis and development."[23]

On the occasion of the contract award to Northrop Grumman, Golden Spike officials revealed a list of partners on the Lunar Lander Systems (LLS) hardware elements. This included Armadillo Aerospace (now dissolved), International Lunar Observatory Association, Masten Space Systems, Moon Express, Paragon Space Development Corp, Southwest Research Institute, Space Florida, United Launch Alliance, and Zero Point Frontiers Corporation.[24]

Shackleton Energy Company

The Shackleton Energy Company,[25] which was founded in 2007 in Del Valle, Texas, has indicated that it has plans to be the first operational lunar mining enterprise and to do so by 2020.[26] According to Bill Stone, SEC's chairman, "[A]fter a phase of robotic prospecting, our crews will establish the infrastructure in space and base camps in the lunar polar crater regions to supervise industrial machinery for mining, processing and transporting lunar products to market in low Earth orbit (LEO) and beyond. It is essential that we seize this opportunity now to provide low cost propellants in space as a means to jump-start the new space economy."[27]

This effort is in several ways the most controversial of the various U. S. commercial space ventures. The other ventures are focused on near Earth orbits that could be potential future threats to the world if they were to crash down from space. Further, the status of these small bodies as "celestial objects" have not yet been clearly defined. The Moon, as discussed later, is clearly addressed in the Outer Space Treaty and is considered to be a global commons resource by most countries of the world. Finally the resources that SEC would seek to remove from the Moon are considered quite valuable and irreplaceable.

In the past, fraudulent efforts have been undertaken involving the "ownership" of elements of Moon. These have included not very serious and bogus efforts such as "selling titles" to tiny parcels on the Moon. This has been recognized as a sort of whimsical prank that had no real legal consequences other than people getting a piece of paper that could be shown to people as a lark. The Shackleton Energy Company, however, has very serious financial capital behind it from Texas energy concerns and have begun to move forward with actual plans to mine and "expropriate resources" from the Moon.

This private initiative that aspires to the future of exploration and mining of

[23] Chris Bergin, "Golden Spike Contract Northrop Grumman for Lunar Lander" NASA Spaceflight.com (January 3, 2013): http://www.nasaspaceflight.com/2013/01/golden-spike-northrop-grumman-lunar-lander/ (Date accessed: August 20, 2015).

[24] *Ibid.*

[25] http://www.shackletonenergy.com/ (Date accessed: April 2, 2013).

[26] "Energy company wants to be first to mine the moon," http://www.networkworld.com/community/blog/energy-company-wants-be-first-mine-moon (Date accessed: April 2, 2013).

[27] Doug Messier, "Exclusive: Shackleton Energy Company Launches Plan for First Lunar Mining Operation," November 9, 2011: http://www.parabolicarc.com/2011/11/09/exclusive-shackleton-energy-company-launches-plan-for-first-lunar-mining-operation/ (Date accessed: April 2, 2013).

the Moon poses real policy, legal and regulatory issues concerning what activities might be conducted on the Moon and could also pose space security issues and concerns as well. The section that follows on U. S. governmental policy and regulatory issues is most relevant to consider in terms of the plans of Shackleton Energy Company.

Moon Express

Moon Express is a privately funded U. S. commercial space company created to develop the resources of the Moon. The stated aim of this startup company is to "send a series of robotic spacecraft to the Moon for ongoing exploration and commercial development." [28]

Moon Express was founded in August 2010 by Naveen Jain, Barney Pell and Robert D. Richards in Mountain View, California, near NASA Ames Research Center. The company seeks to offer commercial lunar robotic transportation and data services with a long-term goal of mining the Moon for rare earth metals such as yttrium, dysprosium and niobium as well as such elements as helium-3.

On June 30, 2011, the company held its first successful test flight of a prototype lunar lander system called the Lander Test Vehicle (LTV) that was developed in partnership with NASA. Later in 2011, Moon Express also announced that it had set up a robotics lab for a lunar probe named the "Moon Express Robotics Lab for INnovation" (MERLIN). In mid-2012, Moon Express announced that it will work with the International Lunar Observatory Association (ILOA) to put a small astronomical telescope on the Moon. Additional details were released in July 2013, stating that there would be two telescopes: a 2-m radio telescope as well as an optical telescope. The currently preferred location is the 5-km (3.1-mile) in diameter Malapert Crater, with a possible launch date in 2018.[29]

On April 30, 2014, NASA also announced that Moon Express was one of the three companies selected for the Lunar Cargo Transportation and Landing by Soft Touchdown (Lunar CATALYST) program. NASA has now agreed to a 3-year no-funds-exchanged Space Act Agreement (SAA) with Moon Express.[30]

The B612 Foundation

The B612 Foundation is a non-profit charitable organization based in Mountain View, California. The goal of the private foundation is to "build, launch, and operate an infrared space telescope." This telescope is to be launched into orbit around the Sun in an orbit similar to that of Venus. The prime objective is to find and track threatening asteroids that could crash into Earth.[31] The foundation has signed a contract with Ball Aerospace for its Sentinel

[28] http://www.moonexpress.com/#missions.

[29] Moon Express Announces First Successful Flight Test of Lunar Lander System Developed with NASA Partnership: http://www.spaceref.com/news/viewpr.html?pid=33991 (Date accessed: August 28, 2015).

[30] About Lunar Catalyst http://www.nasa.gov/lunarcatalyst/#.U2RcWaIXJDy.

[31] http://b612foundation.org/about-us/ (Date accessed: April 2, 2013).

Space Telescope,[32] which will possibly be launched by 2018, for mapping and discovery of asteroids. "Mapping the great unknown of the inner Solar System is the first step in protecting Earth from asteroid impacts and in opening up this next frontier." The cost of this program is on the order of $750 million with about 40 % of the funds raised to support this ambitious private undertaking. This project is significant in that it could track and detect NEAs down to 30 m in size and help to create a detailed mapping of potentially hazardous asteroids projecting possible impact orbits a hundred years into the future. This asteroid mapping would identify bodies much smaller than the current guidelines provided to NASA by the U. S. Congress of mapping all potentially hazardous asteroids down to 140 m in size. It is significant in that asteroids of 30 m in size are potential city killers, and the smaller asteroids are orders of magnitude more numerous than asteroids that are 140 m in diameter or larger. Although this is not a commercial project its ability to identify and track a large number of previously identified NEAs could be of significant value to space mining ventures.

Policies Concerning Space Mining, Resource Extraction and Space Colonies

The U. S. government and the FAA in particular were asked by Bigelow Aerospace in December 2014 about possible protection of U. S. assets that might be deployed on the Moon in the form of private space habitats/colonies. In a December letter in response to Bigelow's query, the FAA said it could act through its licensing procedures to protect private lunar operations from being interfered with by others. So, if Bigelow had a base on the Moon, other companies licensed by the FAA couldn't set up operations at the same site without permission. In response to that George Nield, associate administrator for Commercial Space Transportation at the FAA (FAA-AST) in his December 2014 statement seemed to be saying that the U. S. government would provide some form of official protection for such assets deployed on the Moon.

In subsequent statements in February 2015 at the 18th Annual Commercial Space Transportation Conference that is sponsored by the FAA, Nield stated: ".I want to make clear that the FAA today has responsibility to license launches and reentry and nothing in between…. We are allowed to do something called halo review – called a payload review….. We share the proposal with other government agencies, NASA, Department of Defense, Department of Commerce, to see if there are concerns or issues that the other agencies might have.

"We came to the realization that the federal government probably needs to take a look at its overall regulatory framework," he added when it comes to payloads that might be deployed on the Moon. We need to consider if we "are living up to the Outer Space Treaty obligations we have in terms of overseeing and authorizing the private sector activities of our citizens." He went on to clarify: "We are not talking about property rights. What we are talking about is

[32] See "Ball Aerospace/B612 Foundation Sign Contract for Sentinel Mission," Oct. 30, 2012: http://b612foundation.org/newsroom/press-releases/ball-aerospaceb612-foundation-sign-contract-for-sentinel-mission/ (Date accessed: April 2, 2013).

having the U. S. government have a framework that provides some regulatory certainty for industry."[33]

The issues involving the legal status of space habitats located on the Moon are clearly different from commercial projects that would propose to set up mining operations, but it clearly is a step along the way. The Commercial Spaceflight Federation (CSF) has endorsed the statements Nield has made, but clearly such statements currently have no specific legal status, and until a specific lunar project is proposed it is far from clear exactly how the U. S. government will respond and how the various agencies that are involved (i.e., the FAA, NASA, Commerce, DOD, the State Department and the EPA) will officially respond.

NASA, by means of Space Act Agreements with new space mining startups, new flexible mechanisms such as Lunar Catalyst, and various developmental contract awards are encouraging the various types of new commercial space initiatives as outlined in this chapter. Likewise, the FAA, by means of its congressional authorized mandate under its various authorization acts is seeking to promote new commercial space industries in parallel with NASA.

Conclusions

Undoubtedly, in the near future, several private sector entities both within and outside the United States can be expected to join these pioneering enterprises.[34] They will do their own exploration and prospecting for valuable natural resources in space, but will also be taking advantage of the findings, research and surveys carried out by their respective government space agencies. The description and analysis of the strategies and policies followed by other spacefaring nations in the next chapter will show that exploitation of the resources will soon become inevitable.

[33] http://www.parabolicarc.com/2015/02/07/faa-moves-establish-framework-commercial-lunar-operations/#sthash.08s3icfr.dpuf (Date accessed: August 20, 2015).

[34] Adam Mann, "Tech Billionaires Plan Audacious Mission to Mine Asteroid," 23 April 2012, online: http://www.wired.com/wiredscience/2012/04/planetary-resources-asteroid-mining/ (Last accessed: March 13, 2013).

7 Space Enterprises in Russia and the Former Soviet Union

The former Soviet Union, now Russia, has been one of the most active countries involved in the exploration of space, scientific investigation and practical applications of space. There is a long tradition of making practical applications of space there by deploying communications and broadcasting satellites, launching remote sensing and meteorological satellites, and precision navigation and timing satellites to provide these vital services to one of the most geographical diverse and difficult to serve regions on the planet. This chapter thus covers the many relevant past activities of the U.S.S.R. in terms of how they might lead to future space mining activities.

Soviet space activities have involved exploration of the Moon, Venus, Mars and other planets of the Solar System via satellites equipped to examine the atmosphere and soil contents of the planets as well as to investigate the physics of the Sun. This intensive program has positioned present-day Russia to become one of the leading countries in the area of natural resource harvesting in future years, should this type of activity be sanctioned and encouraged. The program seemingly began with the *Venera 1* space probe launched by the U.S.S.R. on February 12, 1961.[1]

The Venera series of missions were the first to make a survey of Venus, and overall this series of satellites has produced the most systematic information about Venus. Overall these missions, particularly *Venera 4* through *Venera 16*, conducted between 1967 and 1985, have largely confirmed that this very hot and high pressure planet would not likely be a candidate for space mining due to its extreme heat and the intense pressure in its atmosphere.

The first three Venera missions were unsuccessful, but eventually *Venera 4* was able to send detailed data back to Earth. The launch of *Venera 1*[2] on February 12, 1961, was followed by *Venera 2* on November 12, 1965,[3] and

[1] NASA.gov Venera 1 NASA.gov. http://nssdc.gsfc.nasa.gov/nmc/masterCatalog.do?sc=1961-002=A (Accessed August 24, 2015).

[2] Venera 1, Venera 1, NASA.gov http://nssdc.gsfc.nasa.gov/nmc/masterCatalog.do?sc=1961-003A. (Accessed August 24, 2015).

[3] NASA.gov Venera 2 http://nssdc.gsfc.nasa.gov/nmc/masterCatalog.do?sc=1965-091A (Accessed on August 24, 2015).

© Springer International Publishing Switzerland 2017

R.S. Jakhu et al., *Space Mining and Its Regulation*, Astronautical Engineering, DOI 10.1007/978-3-319-39246-2_7

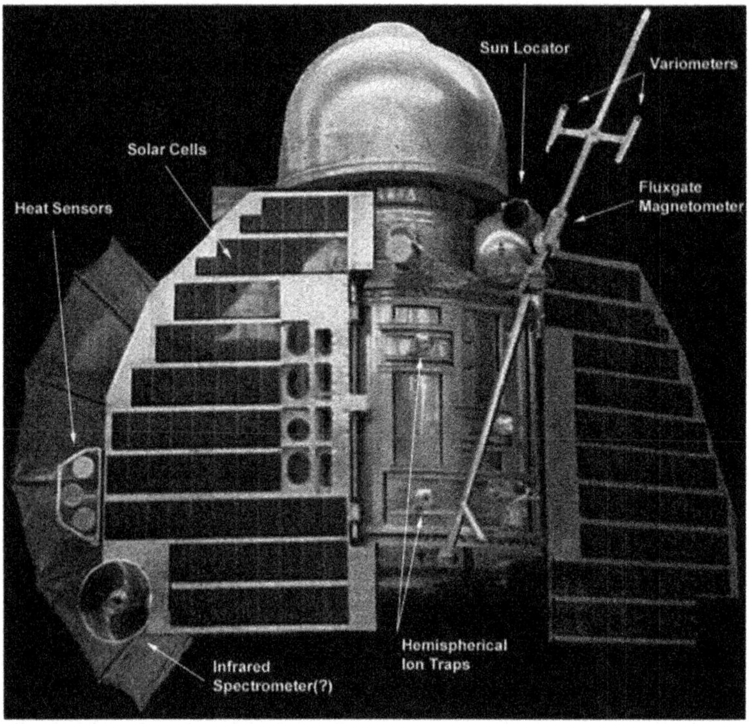

Fig. 7.1 The *Venera 1* space probe that made an early flight to Venus (Graphic courtesy of the Russian Space Agency, Roscomosos.)

then *Venera 3* on November 16, 1965[4] (see Fig. 7.1).

The first successful mission came in June 1967 with the *Venera 4* mission that had the "scientific objective of in-situ studies of the atmosphere of Venus down to the surface."[5] Thus *Venera 4* was "the first probe to transmit data from the atmosphere of another planet." This mission discovered that the Venusian atmosphere is mainly composed of carbon dioxide (between 90 and 95 %). Atomic hydrogen at a 9900 km altitude was found as well.[6] *Venera 5* and *Venera 6* missions were launched in January 1969 "to make parallel measurements as they descended through the Venusian atmosphere."[7]. These missions confirmed "the high temperatures, pressures, and carbon dioxide composition of the atmosphere originally found by *Venera 4*."[8]

[4] NASA.gov Venera 3, NASA http://nssdc.gsfc.nasa.gov/nmc/masterCatalog.do?sc=1965-092A (Accessed August 24, 2015.

[5] NASA.gov Venera 4, http://nssdc.gsfc.nasa.gov/nmc/masterCatalog.do?sc=1967-058A (Accessed on August 24, 2015).

[6] *Ibid.*

[7] NASA.gov Venera 5, NASA http://nssdc.gsfc.nasa.gov/nmc/masterCatalog.do?sc=1969-001A. See also Venera 6, NASA http://nssdc.gsfc.nasa.gov/nmc/masterCatalog.do?sc=1969-002A (Accessed August 24, 2015).

[8] *Ibid.*

The first spacecraft to return data after landing on another planet was *Venera 7* launched in August 1970 with the objectives "to return data from the Venus atmosphere, make a landing on the surface, and continue to return data after landing."[9] Data from *Venera 7* confirmed that the surface temperature of Venus was 237–246 ° C. *Venera 8* was launched in March 1972 "to make a more sophisticated set of scientific measurements at the Venus surface, including studies of the Venus regolith." The *Venera 8* mission confirmed "the earlier data on the high Venus surface temperature and pressure (470 ° C, 90 atmospheres) returned by *Venera 7*." *Venera 8* returned the "first measurements of the surface regolith of Venus" and made "a profile of the cloud layer, including detection of sulfuric acid." The *Venera 9* descent craft/lander was launched in June 1975 with the scientific objectives to make "measurements in the Venus atmosphere and on the surface." The *Venera 10* spacecraft was identical *Venera 9* and launched in the same month, with the scientific objectives to "make in-situ measurements of the Venus atmosphere and surface." *Venera 9* was the first spacecraft to send back images from the surface of another planet. Preliminary results from the *Venera* 9 mission indicated atmospheric constituents including HCl, HF, Br, and I.[10]

Venera 11 and *Venera 12* were identical spacecraft. These were launched in September 1978 "to study the detailed chemical composition of the atmosphere, the nature of the clouds, the thermal balance of the atmosphere, and the composition and mechanical properties of the surface."[11] The two missions discovered, among other things, that there was sulfur and chlorine in the cloud layers and carbon monoxide at low altitudes. *Venera 13* and *Venera 14,* also identical spacecraft, were launched in October and November of 1981, respectively, and were also funded by the Soviet Academy of Sciences to make "studies of the Venus atmosphere and surface." *Venera 13* revealed through the first color images of the surface of Venus that the planet has "an orange-brown flat bedrock surface covered with loose regolith and small flat thin angular rocks."[12] The composition of the sample was classified as "weakly differentiated melanocratic alkaline gabbroids, similar to terrestrial leucitic basalt, with a high potassium content." In addition to that, the *Venera 14* mission "identified three distinct cloud layers."[13] The final two satellites in the initial Venera series were launched in June 1983. The *Venera 15*

[9] NASA.gov Venera 7, http://nssdc.gsfc.nasa.gov/nmc/masterCatalog.do?sc=1970-060A. (Accessed on August 24, 2015).

[10] NASA.gov Venera 9 Descent Craft, NASA http://nssdc.gsfc.nasa.gov/nmc/masterCatalog.do?sc=1975-050D (Accessed on August 24, 2015).

[11] NASA.gov Venera 11 Descent Craft, NASA http://nssdc.gsfc.nasa.gov/nmc/masterCatalog.do?sc=1978-084D (Accessed on August 24, 2015).

[12] NASA.gov: Venera 13 Descent Craft, NASA http://nssdc.gsfc.nasa.gov/nmc/masterCatalog.do?sc=1981-106D(Accessed on August 24, 2015) Also see: Venera 14 Descent Craft, NASA http://nssdc.gsfc.nasa.gov/nmc/masterCatalog.do?sc=1981-110D. (Accessed on August 24, 2015).

[13] NASA.gov Venera 14 Descent Craft, NASA http://nssdc.gsfc.nasa.gov/nmc/masterCatalog.do?sc=1981-110D. (Accessed on August 24, 2015).

and *16* mission objective was "to study the surface properties of Venus." *Venera 15* and *Venera 16* "produced a map of the northern hemisphere [of Venus] from the pole to 30°N." This map enabled scientists to find "several hot spots, possibly caused by volcanic activity," on the surface of Venus.[14]

Future missions under the Venera series are the Venera-D and the Venera-Glob projects that are planned for a 2021–2024 landing. Venera-D and Venera-Glob's funding and development was not expected to start until the whole new revision of the Russian space program is approved by the Russian government for the 2016–2025 period.

These projects could include a radar-carrying orbiter, several small surface landers and aerial vehicles. One of the landers could be designed to extend the survival time on the surface achieved by its predecessors. Scientists also are considering deploying atmospheric balloons at various altitudes in the atmosphere of Venus for more than a month-long mission. In turn, balloons could release their own mini-probes. Finally, a special wind-flying aircraft or a glider, originally considered for the Venera-D project, could finally fly with Venera-Glob. Project planners also considered a real-time interaction between the Venera-Glob and Venera-D projects.[15]

The former Soviet Union also had the *Vega 1* and *2* projects in addition to the *Venera 1* through *16*. These Vega projects were carried out in cooperation with Austria, Bulgaria, Hungary, former East Germany, former West Germany, Poland, former Czechoslovakia, and France. The spacecraft were launched in December 1984, *Vega 1* on December 15 and *Vega 2* on December 21. The Vega mission had "three major goals: to place advanced lander modules on the surface of Venus, to deploy balloons (two each) in the Venusian atmosphere, and, by using Venusian gravity, to fly the remaining buses past the Comet Halley." Although the "*Vega 1* soil experiment failed," the *Vega 2* lander collected and investigated "a soil sample; the experiment identified an anorthosite-troctolite rock—rarely found on Earth but present in the lunar highlands."[16]

The former Soviet Union also undertook a number of exploratory missions to gather data on Mars. In July 1988 it launched the Phobos mission, named after one of Mars's two small moons. The purpose of this Mars mission was to "characterize the plasma environment in the Martian vicinity," "conduct surface and atmospheric studies….and to study the surface composition of Phobos as well." The mission involved two probes: *Phobos 1* (launched on July 7) and *Phobos 2* (launched on July 12). Although *Phobos 1* failed to reach Mars, *Phobos 2* succeeded and collected "data on Sun irradiation, the interplanetary medium between Earth and Mars, as well as some data on both Mars, and

[14] NASA.gov Venera 15, NASA http://nssdc.gsfc.nasa.gov/nmc/masterCatalog.do?sc=1983-053A. (Accessed on August 24, 2015).

[15] Zasova, L., Venera: Izuchenie prodolozhaetsya, 10.03.2011, Accessed on June 15, 2011, at: http://www.venera-d.cosmos.ru/index.php?id=692&tx_ttnews%5btt_news%5d=1288&cHash=f9bfd2c6e7616171412b316d206d73a4 (Accessed August 25, 2015).

[16] NASA.gov Vega 1, Solar System Exploration, http://solarsystem.nasa.gov/missions/profile.cfm?MCode=Vega_01&Display=ReadMore (Accessed August 25, 2015).

Phobos." The *Phobos 2* mission ended after 8 months on March 27, 1989.[17]

Russia's follow-up Phobos mission to Mars, named the Phobos-Grunt mission, failed. The objective of the mission was to obtain soil samples from the Martian satellite Phobos and return them to Earth." Following the failure of the Phobos-Grunt mission, a new mission with the same objective, named the Phobos-Grunt 2 mission, is pending approval in the 10-year plan for the 2016–2025 period. The Russian Federal Space Agency [Roscosmos] and the Russian Academy of Sciences have already approved the project, and if funded it would fly in 2025.[18]

In addition to the Phobos mission, the former Soviet Union launched a number of additional Mars probes. Of these launches, the successful or partially successful missions included *Mars 1* (launched November 1, 1962), *Mars 2* (launched May 19, 1971), *Mars 3* (launched May 28, 1971), *Mars 5* (launched July 25, 1973) and *Mars 6* (launched August 5, 1973). The other missions experienced launch or other types of failure.[19]

The *Mars 1* mission was launched "with the intent of flying by the planet [Mars] at a distance of about 11,000 km." It was designed to "image the surface and send back data on…atmospheric structure, and possible organic compounds." However, the closest approach of *Mars 1* to Mars occurred on June 19, 1963, at a distance of approximately 193,000 km. It was nevertheless able to send back useful data. The *Mars 2* and *Mars 3* missions were identical craft and launched to study Mars from the surface. Although the *Mars 2* lander crashed, it was "the first human-made object to make contact with Mars." The *Mars 3* lander made the "first survivable landing on Mars." The two probes "sent back a total of 60 pictures." Images and data obtained from these two probes revealed: "atomic hydrogen and oxygen in the upper atmosphere, [and]… water vapor concentrations 5000 times less than in Earth's atmosphere."[20]

Mars 4, 5, 6 and *7* "comprised an associated group of Soviet spacecraft launched towards Mars in July and August of 1973. These spacecraft were designed to "orbit Mars and return information on the composition, structure, and properties of the Martian atmosphere and surface." The *Mars 5* mission discovered "U, Th, and K composition similar to terrestrial mafic rocks." *Mars 6,* however, was designed as a lander. Key findings from these missions included the following: "[a] high water vapor content (100 precipitable μm)" was present south of Tharsis region, plus an ozone layer…at 40 km altitude with about one-thousandth the concentration of Earth's." Images of the Martian surface suggested "erosion caused by flowing water." The *Mars 6* craft that was a lander revealed "several

[17] Phobos (1988–1989), Journey through the Galaxy, Department of Astronomy, Case Western Reserve University http://burro.astr.cwru.edu/stu/advanced/20th_soviet_phobos.html. (Accessed August 25, 2015)

[18] Phobos-Grunt-2, RussianSpaceWeb.com http://www.russianspaceweb.com/phobos_grunt2.html (Accessed August 25, 2015).

[19] Mars 3 Lander, NASA http://nssdc.gsfc.nasa.gov/nmc/spacecraftDisplay.do?id=1971-049F.

[20] Mars (1960–1974), Journey through the Galaxy, Department of Astronomy, Case Western Reserve University http://burro.astr.cwru.edu/stu/advanced/20th_soviet_mars.html.

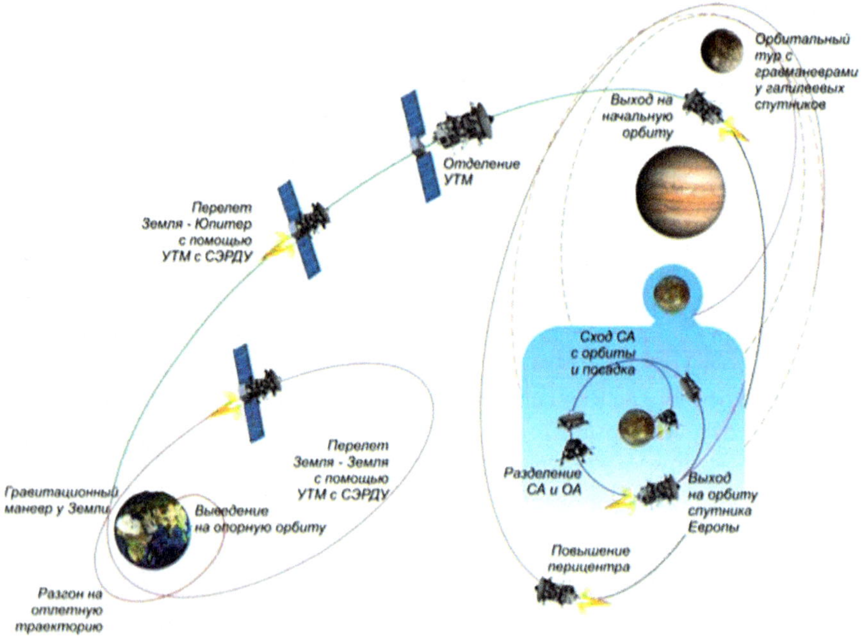

Fig. 7.2 Flight profile of the mission to Ganymede with a separate lander (Graphic courtesy of the Russian Space Agency, Roscosmos.)

times more atmospheric water vapor than previously reported" and "an estimate of argon abundance in the atmosphere of 25–45 %."[21]

Until recently, there had not been any Russian or Soviet mission beyond Mars. The first mission that would go beyond Mars would be the Laplas-P mission. This planned future mission would study Jupiter's moon Ganymede.[22] (see Fig. 7.2).

In the original conception of 2009 this was to have been an exploratory mission to Jupiter's moon Europa. The mission re-envisioned is now planned to go to Jupiter as an orbiter and then have a separate lander that would touch down on Ganymede. This mission is currently planned for 2023 and is to be launched by either a Proton or an Angara launcher.[23]

The Soviet Union space exploration program has concentrated on Venus, Mars and the Moon. Of the three, the greatest amount of effort has been directed at the Moon with the Luna series 1–24 returning a great deal of information. There are additional missions now planned as described below.

[21] Mars 6, Solar System Exploration, NASA https://solarsystem.nasa.gov/missions/profile.cfm?Sort=Alpha&Letter=M&Alias=Mars%206.

[22] Russia funds a proposal to land on Jupiter's moon Ganymede, RussianSpaceWeb.com http://russianspaceweb.com/laplas.html.

[23] Russian Funds a Proposal to Land on Jupiter's Moon Ganymede, http://www.russianspaceweb.com/laplas.html (Accessed August 25, 2015).

To date 24 lunar missions have been launched by the Soviet Union and Russia under the Luna series of missions. The following reports give highlights from these various missions. The Soviet Union's first successful lunar mission was the *Luna 1* mission. *Luna 1* was the first spacecraft to escape Earth's gravity, to enter into orbit around the Sun, and to fly past the Moon. *Luna 1* was launched in January 1959 to impact the Moon. The mission was partially successful in that it failed to impact the Moon but, instead, flew past the Moon. *Luna 2* was launched in September 1959 and accomplished what *Luna 1* had intended—namely it became the first spacecraft to reach the surface of the Moon. The sensors of the *Luna 2* spacecraft did not find any evidence of a lunar magnetic field or radiation belt.[24]

The *Luna 3* mission was launched in October 1959. Its objective was "to take the first photographs of the far side of the Moon." The images returned by the *Luna 3* spacecraft covered some 70 % of surface of the lunar far side. This data was "used to create the first rudimentary atlas of the previously unrevealed lunar surface." The *Luna 9* spacecraft made "the first survivable landing on the Moon" and was "the first to send back images from the surface of another world." The mission discovered that "a lander would not sink into a thick layer of dust."[25] The first spacecraft to orbit the Moon was the *Luna 10* spacecraft, and it was able to provide a continuous study of the lunar surface. The spacecraft conducted "extensive research in lunar orbit, gathering important data on the nature of lunar rocks (which were found to be comparable to terrestrial basalt rocks).",[26]

The *Luna 11* mission was launched in August 1966, and it was "designed to take the Soviet Union's first detailed pictures of the surface of the Moon from lunar orbit and to obtain data on the Moon's composition and confirm the mass concentrations first detected by *Luna 10*." Although the orbiter successfully entered lunar orbit, the failure of the TV system to return usable images resulted in limited success of the mission. The *Luna 12* mission was launched in October 1966 "to complete the mission that *Luna 11* had failed to accomplish. This was to take high-resolution photos of the Moon's surface from lunar orbit." The spacecraft successfully photographed "part of the lunar surface at a resolution down to 15–20 m, the best to date."[27]

---

[24] Luna 1, Solar System Exploration, NASA.gov https://solarsystem.nasa.gov/missions/profile.cfm?Sort=Target&Target=Moon&MCode=Luna_01. Also see

Luna 2, Solar System Exploration, NASA.gov https://solarsystem.nasa.gov/missions/profile.cfm?Sort=Target&Target=Moon&MCode=Luna_02 (Accessed August 25, 2015).

[25] Luna 9, Solar System Exploration, NASA (Accessed August 25, 2015) https://solarsystem.nasa.gov/missions/profile.cfm?Sort=Target&Target=Moon&MCode=Luna_09.

[26] Luna 10, Solar System Exploration, NASA.gov (Accessed August 25, 2015) https://solarsystem.nasa.gov/missions/profile.cfm?Sort=Target&Target=Moon&MCode=Luna_10&Display=ReadMore.

[27] Luna 11, Solar System Exploration, NASA.gov (Accessed August 25, 2015) https://solarsystem.nasa.gov/missions/profile.cfm?Sort=Target&Target=Moon&MCode=Luna_11. Also see

Luna 12, Solar System Exploration, NASA.gov (Accessed August 25, 2015) https://solarsystem.nasa.gov/missions/profile.cfm?Sort=Target&Target=Moon&MCode=Luna_12&Display=ReadMore.

Luna 13 was launched in December 1966 to "land on the Moon and characterize the lunar surface." The lander probe returned "5 panoramic photos of the landscape at different Sun angles" and measured the lunar "soil's physical and mechanical properties." *Luna 14* was launched in April 1968 with the primary objective to "test communications systems in support of the Soviet effort to send cosmonauts to the Moon." The mission met its primary objective and, additionally, provided "data for studies of the lunar gravitational field, the solar wind, cosmic rays, lunar motion, and the interaction of the Earth and lunar masses."[28]

The *Luna 16* mission was launched in September 1970 "to drill out a sample core from the Moon's surface and return it to Earth." The spacecraft successfully returned samples to Earth and became "the first fully robotic sample return of any nation." It was found that the dark, powdery basalt material was "very similar to that obtained from another mare site by Apollo 12" but "differed slightly from *Apollo 11*'s samples in the levels of titanium and zirconium oxide."[29]

Luna 17 was also known as the *Lunokhod 1* mission. It was launched in November 1970 to "deliver a robotic rover called 'Lunokhod 1' to the lunar surface." The mission of the rover was to "travel to various locations under the real-time control of operators on Earth and conduct tests on the lunar soil." The *Lunokhod 1* rover was the first wheeled unmanned robotic spacecraft to drive on another world." This mission was able to perform 25 soil analyses and tested the "soil's mechanical characteristics at more than 500 locations."[30]

Luna 18 was a failure, but in September 1971 the *Luna 19* mission was successfully placed in orbit around the Moon and began collecting data about the Moon and its terrain and composition. The *Luna 19* spacecraft "orbited the Moon for one year, transmitting photos of the Moon and data on the composition of the lunar surface."

The *Luna 20* mission was launched in February 1972 "to conduct the mission that *Luna 18* was unable to complete. The objective was to obtain a soil sample from the lunar highlands and bring it back to Earth for comparison with *Luna 16*'s sample from a mare site." The mission successfully returned a 55-g soil sample and discovered pure iron. It was revealed that the returned "sample differed from that collected by *Luna 16* in that the majority (50–60 %) of the rock particles in the newer sample were ancient anorthosite (which consists largely of feldspar) rather than the basalt of the

[28] Luna 13, Solar System Exploration, NASA.gov (Accessed August 25, 2015) https://solarsystem.nasa.gov/missions/profile.cfm?Sort=Target&Target=Moon&MCode=Luna_13. Also see:

Luna 14, Solar System Exploration, NASA.gov (Accessed August 25, 2015) https://solarsystem.nasa.gov/missions/profile.cfm?Sort=Target&Target=Moon&MCode=Luna_14.

[29] Luna 16, Solar System Exploration, NASA.gov (Accessed August 25, 2015) https://solarsystem.nasa.gov/missions/profile.cfm?Sort=Target&Target=Moon&MCode=Luna_16.

[30] Luna17/Lunokhod 1, Solar System Exploration, NASA.gov (Accessed August 25, 2015) https://solarsystem.nasa.gov/missions/profile.cfm?Sort=Target&Target=Moon&MCode=Luna_17.

earlier one (which contained about 1–2 % of anorthosite)." [31]

The *Luna 21/Lunokhod 2* mission was launched in January 1973. Its objective was to deliver safely the second Soviet lunar rover, the *Lunokhod 2,* to the Moon. On this mission the objective was to "collect images of the lunar surface…and study mechanical properties of the lunar soil." Among others accomplishments, the rover "covered 37 km of terrain…and conducted at least 740 mechanical tests of the soil."[32]

The *Luna 22* mission was launched in May 1974 with the goal of "orbiting the Moon, imaging the surface and conducting remote-sensing measurements." It was also going to "study the chemical composition of the lunar surface, record meteoroid activity, search for a lunar magnetic field, measure solar and cosmic radiation flux, and continue seeking evidence of an irregular magnetic field." The last Soviet or Russian probe to the Moon was the *Luna 24* spacecraft, which was launched in August 1976. The goal of *Luna 24* was to "land at the site of a large mass concentration on the Moon, obtain a soil sample down to about 2 m beneath the lunar surface, and return it to Earth." On this mission 170.1 g of soil was successfully recovered from the probe. Subsequent study of the sample indicated "a laminated type structure, as if laid down in successive deposits"[33] (see Fig. 7.3).

Future lunar missions under the Luna series were considered by Roscosmos and are now firmly planned albeit with a new almost 3-year delay. This new Russian exploration program will include five probes and are: *Luna-25*, or the Luna-Glob lander (lander mission, with an expected launch date of 2017), *Luna-26*, or Luna-Glob orbiter (orbiter mission, expected launch date 2018), *Luna-27*, or Luna-Resurs lander (lander mission, expected launch date 2019), *Luna-28*, or the Luna-Grunt rover (lunar sample return mission, expected launch date 2025), and *Luna-29*, or the Luna-Grunt sample return vehicle (lunar sample return mission, expected launch date 2025).[34] These missions will provide specific data on the mineral content of the Moon and could provide useful data to any enterprise that would contemplate mining on the lunar surface.

The first of these missions is now scheduled for 2017, as is depicted in Fig. 7.4. This will be the first lander mission since *Luna 24,* which that took place in 1976.[35]

In addition to the extensive Luna exploratory missions to the Moon, there was also the Zond program, which included eight

[31] Luna 19, Solar System Exploration, NASA.gov (Accessed August 25, 2015) https://solarsystem.nasa.gov/missions/profile.cfm?Sort=Target&Target=Moon&MCode=Luna_19. Also see:

Luna 20, Solar System Exploration, NASA.gov (Accessed August 25, 2015) https://solarsystem.nasa.gov/missions/profile.cfm?Sort=Target&Target=Moon&MCode=Luna_20.

[32] Luna 21/Lunokhod 2, Solar System Exploration, NASA.gov (Accessed August 25, 2015) https://solarsystem.nasa.gov/missions/profile.cfm?Sort=Target&Target=Moon&MCode=Luna_21.

[33] Luna 22, Solar System Exploration, NASA.gov (Accessed August 25, 2015) https://solarsystem.nasa.gov/missions/profile.cfm?Sort=Target&Target=Moon&MCode=Luna_22.

[34] Russian Moon missions face three-year delay, RussianSpaceWeb.com http://www.russianspaceweb.com/spacecraft_planetary_2014.html.

[35] Luna Glob Mission (Accessed August 25, 2015) http://www.russianspaceweb.com/images/spacecraft/planetary/moon/luna_glob/lg_lander_2011_1.jpg.

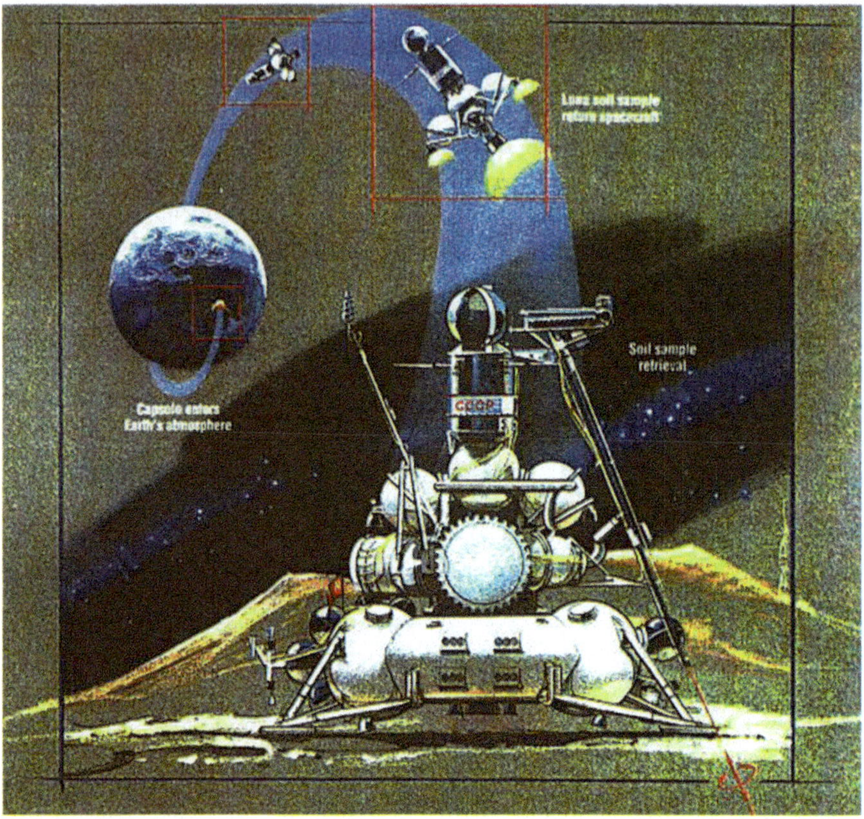

Fig. 7.3 The 1976 Soviet *Lunar 24* Moon mission that provided a soil sample return (Graphic courtesy of the Russian Space Agency, Roscosmos.)

Zond probes that were launched in the mid-1960–1970. These spacecraft also went to Venus and Mars as well as the Moon; the missions were largely flyby and orbital missions that took high resolution images. The most significant part of the Zond missions to the Moon was their ability to complete the other 30 % mapping of the far side of the Moon.

The first fully successful Soviet mission to fly around the Moon and land safely back on Earth was the *Zond 7* mission that took color photographs of the Moon. With similar objectives like its predecessors, the *Zond 8* spacecraft "successfully looped around the Moon and returned to Earth" while taking high-quality of photos of the Moon and conducting measurements of the lunar environment. However, none of the Zond missions were designed to search for lunar natural resources and, consequently, discovered little that was of significance as to the possible location of natural resources in space.[36]

Currently it is not clear whether there are any specific plans for the Russian

[36] Zond 7, Solar System Exploration, NASA.gov (Accessed August 25, 2015) https://solarsystem.nasa.gov/missions/profile.cfm?Sort=Target&Target=Moon&MCode=Zond_07.

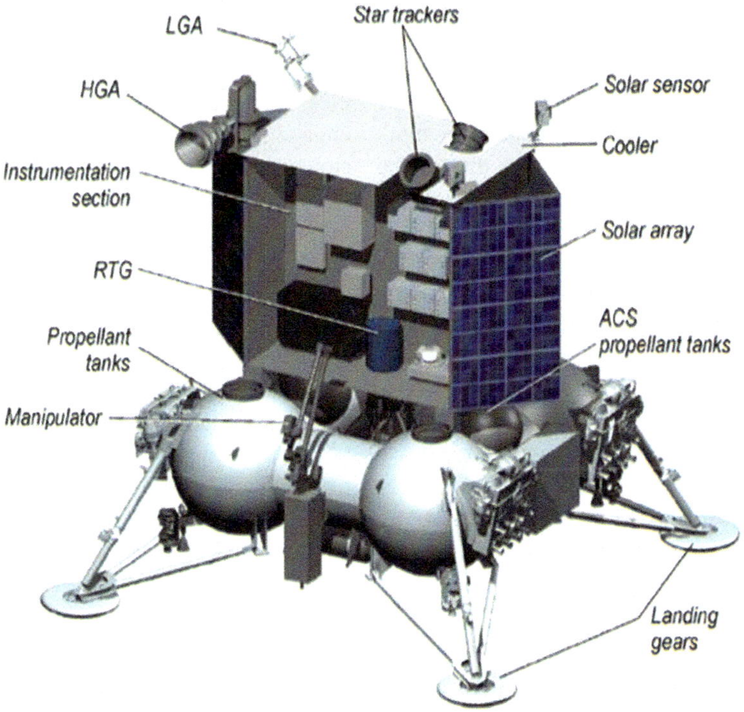

Fig. 7.4 Russian *Luna 25* (or Luna-Glob) lander that will provide updated soil analysis (Graphic courtesy of the Russian Space Agency, Roscosmos.)

government or any Russian entity to engage in space mining, but in terms of experience and demonstrated past capabilities, it seems clear that there is very extensive Russian capability and the planned *Luna 25* through *Luna 29* missions suggests a continued strong interest in the Moon and its mineral content. One of the partners in the Russian Luna missions, namely the Sternberg Institute, has indicated that there is particular interest in exploring the lunar surface to see if it is rich in rare earth metals. The head of the Department of Lunar and Planetary Research at the Sternberg Institute, Vladislav Shevchenko, says Moon exploration could be the solution to the current shortage of rare earth metals, whose production is controlled by China. Shevchenko has suggested that deliveries from space could potentially be more cost-effective than mining for metals at home.

Conclusions

The Russian government is, however, initiating a new rare earth's metals mining operation in the Urals as perhaps a more relevant and nearer term response to the immediate needs for these precious metals. Clearly Russia's proposed investment in five new space probes going to the Moon that involves an estimated cost of $2.5 billion represents a clear interest in the Moon that may go beyond just scientific inquiry.[37]

The U. S. Congress now has under consideration the so-called "Asteroid Act" that would serve to protect the mining claims of U. S. entities that might engage in space mining, and it includes a "reciprocity clause" that would allow countries such as the United States and Russia, the United States and China, or the United States and Europe to mutually accept the claims of other countries in exchange for recognition of U. S. claims. Today only a relatively small number of countries (less than ten)—or their nationally regulated businesses—could possibly take on an enterprise as daunting as space mining. Thus reciprocity might be one way to address this issue. This does not, however, address the prohibition in the Outer Space Treaty of "appropriating" the Moon or other celestial objects.[38, 39]

A number of other countries, as we shall see in later chapters, might have an interest in space mining. This interest could be manifested in a variety of ways in the future. These countries, or companies from these countries, might not engage in a complete space mining mission themselves but might seek partnerships with those planning such initiatives. Planetary Resources and Deep Space Industries, in particular, have been active in seeking partners with special expertise in key technologies.

Some of the countries that are quite experienced in mining and mining technology and systems such as Australia, Canada, and Brazil have also developed some expertise in the space systems and commercial space transportation. Currently the leading developers of automated mining are in western Australia mines, where virtually every element of mining operations are automated or involve remote tele-operation of robotic and transport systems.

Russian interest in space mining, along with many other countries, clearly has reason to be strong. Yet until the regulatory issues surrounding this field are resolved the way forward remains clouded. Clearly Russia will monitor relevant efforts around the world and note progress in related space technology, remote mining and more pending efforts to clarify how space mining might be accomplished and under what regulatory regime.

---

[37] Cecilia Jasasmie, "Russia Pushes Forward Plans to Mine the Moon", Mining.com http://www.mining.com/wp-content/uploads/2014/10/russia-is-seriously-advancing-plans-to-mine-the-moon.jpg.

[38] Adam Minter, "The Asteroid-Ming Race Begins" Bloomberg View, September 8, 2014, http://www.bloombergview.com/articles/2014-09-08/the-asteroid-mining-race-begins.

[39] Adam Minter, "The Asteroid-Ming Race Begins" Bloomberg View, September 8, 2014, http://www.bloombergview.com/articles/2014-09-08/the-asteroid-mining-race-begins.

8 Activities in Europe, Canada and Other Western Countries

Europe has been engaged in a number of exploratory space missions to study not only planetary objects within the Solar System but also the chemical and physical composition of the broader universe, especially with initiatives that have started within the last 15 years. These activities have been driven by many concerns, which include a basic interest in astrophysics, a better understanding of cosmic threats and potential planetary defense, and even practical future applications that might in the time involve such activities as planetary mining.

Herschel Space Observatory

On May 14, 2009, ESA launched the Herschel space observatory "to study the formation of stars and galaxies, and to investigate the relationship between the two."[1] This space observatory is the first to detect water vapor in "a molecular cloud on the verge of star formation." It detected the equivalent more than "2000 Earth oceans-worth of water vapor … liberated from icy dust grains by high-energy cosmic rays passing through the cloud."[2]

The Herschel space observatory also has made a number of other significant and even startling findings. These included finding "the first robust evidence of molecular oxygen in the Orion Nebula." It was particularly noteworthy that the observed profusion was "ten times larger than indicated by previous observations of other molecular clouds."[3] The space observatory also identified the first comet (Comet 103P/Hartley 2) that actually contains "···water that is similar in isotopic composition to that in Earth's

[1] ESA, Fact Sheet: Herschel, at 1–2, online: ESA http://esamultimedia.esa.int/docs/herschel/Herschel-Factsheet.pdf.

[2] ESA, News, "Large Water Reservoirs at the Dawn of Stellar Birth" (October 9, 2012), online: ESA http://sci.esa.int/herschel/50907-large-water-reservoirs-at-the-dawn-of-stellar-birth.

[3] ESA, News, "Herschel uncovers "hidden" Oxygen in Orion" (August 1, 2011), online: ESA http://sci.esa.int/herschel/49008-herschel-uncovers-hidden-oxygen-in-orion/.

© Springer International Publishing Switzerland 2017

R.S. Jakhu et al., *Space Mining and Its Regulation*,
Astronautical Engineering, DOI 10.1007/978-3-319-39246-2_8

oceans."[4] The observatory also discovered water vapor around Ceres, the largest object in the Asteroid Belt that lies between the orbits of Mars and Jupiter. Further the observatory "discovered that a molecule vital for creating water exists in the burning embers of dying Sun-like stars." Overall these survey results from the Herschel observatory "revealed that the reservoir of molecular gas [and water vapor] in the Milky Way is hugely underestimated—almost by one third—when it is traced with traditional methods."[5]

Mars Express

Mars Express, the first European mission to another planet, was launched in June 2003 to monitor "all aspects of the [Martian] environment." The mission objective was to "study the [Martian] atmosphere and climate, the planet's structure, its mineralogy and its geology, and to search for traces of water."[6] The European Space Agency's Mars Express mission has been closely coordinated with NASA's Mars missions in order to leverage the findings of both exploratory efforts. As part of this coordinative effort is the possibility of extending the mission for an additional year.[7]

Mars Express discovered "a family of minerals that form only in the presence of water that are known as phyllosilicates." In addition certain sulfates and iron-based minerals that form in dry conditions without water were identified. Mars Express was able to observe the north and south polar ice caps of the Red Planet in great detail. These significant observations confirmed that the composition of the polar caps was mostly water-ice and the amount of ice at the south pole could "cover the entire planet with a layer of water 11 m deep." The mission found methane on Mars and high-altitude carbon dioxide ice clouds in the atmosphere of Mars. It revealed that hydrogen and oxygen ions dominate the escape from the atmosphere of Mars, while the escape of carbon oxides are small and that the atmosphere is supersaturated with water vapor.[8]

[4] ESA, News, "Herschel finds First Evidence of Earth-like Water in a Comet" (5 October 2011), online: ESA http://sci.esa.int/herschel/49386-herschel-finds-first-evidence-of-earth-like-water-in-a-comet/.

[5] "Herschel finds First Evidence of Earth-like Water in a Comet" (5 October 2011), online: ESA http://sci.esa.int/herschel/49386-herschel-finds-first-evidence-of-earth-like-water-in-a-comet/.

Also see: ESA News, "New Molecules around Old Stars" (June 17, 2014), online: ESA http://sci.esa.int/herschel/54158-new-molecules-around-old-stars/.

Also see: ESA News, "There is more Gas in the Galaxy than is dreamt of by Astronomers" (June 11, 2013), online: ESA http://sci.esa.int/herschel/51909-there-is-more-gas-in-the-galaxy-than-is-dreamt-of-by-astronomers/. "In the Milky Way, as well as in other galaxies, stars are born from the collapse of the densest and coldest clumps of matter in a molecular cloud. These clouds are gigantic star-forming complexes consisting mainly of molecular hydrogen (H_2), a gas that does not emit any light at the low temperatures found in molecular clouds.

[6] ESA, Mars Express: a Decade of Observing the Red Planet (Berlin: European Space Agency, 2013), online: ESA http://esamultimedia.esa.int/multimedia/publications/BR-312/.

[7] ESA, Mars Express, supra note 330.

[8] Ibid.

Fig. 8.1 ESA's Mars Express depicted above Mars (Graphic courtesy of ESA.)

The specific objectives of *Mars Express* (see Fig. 8.1) and the associated *Beagle 2* lander were as follows:

- Image the entire surface at high resolution (10 m/pixel) and selected areas at super resolution (2 m/pixel).
- Produce a map of the mineral composition of the surface at 100-m resolution.
- Map the composition of the atmosphere and determine its global circulation.
- Determine the structure of the subsurface to a depth of a few km.
- Determine the effect of the atmosphere on the surface.
- Determine the interaction of the atmosphere with the solar wind.

The *Beagle 2* lander was planned to obtain more detailed information on the surface of Mars such as:

- Determine the geology and the mineral and chemical composition of the landing site.
- Search for life signatures (exobiology).
- Study the weather and climate.

As a follow up to the *Mars Express* detailed data collection ESA has now "established the ExoMars program to investigate more closely the Martian environment as a prelude to a future Mars sample return mission in the 2020's." Within this future cooperative program with Russia, two missions are foreseen to be carried out with Roscosmos: "one consisting of an Orbiter plus an Entry, Descent and Landing Demonstrator Module, to be launched in 2016, and the other, featuring a rover, with a launch date of 2018."[9]

Venus Express

Venus Express was Europe's first spacecraft to visit Venus. It was launched by ESA in November 2005 to explore the Venusian atmosphere. The science objectives of the mission were formally stated to be "to study the atmosphere, the plasma environment, and the surface of Venus in great detail." Top discoveries of the mission include "a surprisingly cold region high in the planet's atmosphere, where conditions may be frigid enough for carbon dioxide to freeze out as ice or snow." *Venus Express* also detected a thin layer of ozone and evidence that a significant amount of water had been released into space as water vapor over eons of time. Finally *Venus Express* detected that "deuterium, a heavy form of hydrogen, is progressively enriched and also exists in the upper regions of Venus's atmosphere, as well as a layer of [sulfur] dioxide that lies above the main cloud layer."[10] Proposals for a new Venus mission are currently pending at ESA but not yet formally approved.

Mercury Mission

Europe's first mission to Mercury is known as BepiColombo and is named after Professor Bepi Colombo who was a mathematician and scientist at the University of Padua and an expert in the study of Mercury. This mission is now planned to be launched in January 2017, which is a slippage from the original launch date of July 2016. It is not expected to arrive at Mercury until mid-year 2024. This is a joint mission between ESA and the Japanese Aerospace Exploration Agency (JAXA).

The mission is designed to "provide the measurements necessary to study and understand the composition, geophysics, atmosphere, magnetosphere and history of Mercury." BepiColombo is an ambitious mission comprising two separate orbiters, the ESA-led Mercury Planetary Orbiter (MPO) and the JAXA-led Mercury Magnetospheric Orbiter (MMO), as well as a carrier spacecraft, known as the Mercury Transfer Module (MTM). In addition to carrying out experiments with regard to Einstein's theory of relativity it is intended to investigate polar deposits, including their composition and origin.[11]

Jupiter Exploratory Mission

ESA has planned to launch JUpiter ICy moons Explorer (JUICE) mission in 2022 to make detailed observations of

[9] Online: The ExoMars Programme 2016-2018, Robotic Exploration of Mars, ESA http://exploration.esa.int/mars/46048-programme-overview/.

[10] Venus Express: Mission Overview, Venus Express, ESA http://sci.esa.int/venus-express/33010-summary/.

[11] Fact Sheet, BepiColombo, ESA http://sci.esa.int/bepicolombo/47346-fact-sheet/.

Also see: BepiColombo Launch Moved to 2017/http://sci.esa.int/bepicolombo/55693-bepicolombo-launch-moved-to-2017/.

Jupiter and three of its largest moons: Ganymede, Callisto and Europa. This is a major technical challenge not only to deploy an observatory so as to make these observations, but to withstand the heavy radiation that is found at the closest moon, Europa.

JUICE is the first large-scale endeavor selected as a prime mission for ESA's Cosmic Vision 2015–2025 program. It will be launched from Europe's spaceport in Kourou, French Guiana, on an Ariane 5 rocket with the objective of getting to Jupiter in 2030. It will then spend at least 3 years making detailed observations of Jupiter and especially of Jupiter's diverse Galilean moons, icy Europa and rock-ice Ganymede and Callisto. These moons make the Jovian system a miniature solar system in its own right.[12]

The reason that Europa, Ganymede and Callisto were selected is because all three are thought to host internal oceans and as such are potential habitats for life.

JUICE will continuously observe Jupiter's atmosphere and magnetosphere, and the interaction of the Galilean moons with the gas giant planet. It will survey and obtain measurements of Callisto, the most heavily cratered object in the Solar System. It will twice swing by Europa and make the first measurements of the thickness of Europa's icy crust and will identify candidate sites for future exploration and possible landing. Finally the spacecraft will enter orbit around Ganymede in 2032, where it will study the icy surface and internal structure of the moon.[13]

[12]ESA News, JUICE, ESA http://sci.esa.int/juice/.

[13]"JUICE is Europe's next large science mission," ESA Science and Technology, May 2, 2012. http://sci.esa.int/cosmic-vision/50321-juice-is-europe-s-next-large-science-mission/

All of this will be useful scientifically, but the impulse levels required to reach Jupiter are such that the Jovian moons are not likely candidates for commercial space mining.

European Mission to the Moon

The first European mission to the Moon was the Small Missions for Advanced Research and Technology-1 (SMART-1), launched in September 2003 and ended in September 2006 (see Fig. 8.2). The mission objective was to test and prove "an ion drive and miniaturized instruments, along with investigations of the lunar geochemistry and a search for ice at the south lunar pole." The SMART-1 mission studied the Moon and gathered data "about the morphology and mineralogical composition of the surface in visible, infrared and X-ray light." Although this was important from a scientific viewpoint it also provided useful data with regard to possible mining activities on the lunar surface as well as excavation below.[14]

Over a 16-month period the SMART-1 sensors accumulated a large amount of data that was analyzed after the project ended by being deliberately crashed into the lunar surface. Bernard Foing, ESA SMART-1 Project Scientist, has said that the data collected by SMART-1. "call into question the theories concerning the Moon's violent origin and evolution." He noted that the data suggested that the Moon may have formed from the impact of a Mars-size asteroid with Earth 4500 million years

[14]Fact Sheet, Smart-1, ESA http://sci.esa.int/smart-1/47367-fact-sheet/.

Fig. 8.2 SMART-1 trajectory around the Moon up until it crashed (Graphic courtesy of ESA.)

ago. "SMART-1 has mapped large and small impact craters, studied the volcanic and tectonic processes that shaped the Moon, unveiled the mysterious poles, and investigated sites for future exploration."

"ESA's decision to extend the SMART-1 scientific mission for a full 16 months—or 10 months more than first planned—allowed the instrument scientists to extensively use a number of innovative observing modes. The initial major goal of the mission was to test an ion engine (solar electric propulsion) in space for the first time for interplanetary travel, and to allow the capture of a spacecraft and to put it into orbit around another celestial body, in combination with gravity assist maneuvers.

SMART-1 also tested future deep-space communication techniques for spacecraft, techniques to achieve autonomous spacecraft navigation, and miniaturized scientific instruments, used for the first time around the Moon. [15]

ESA's Deep Space Missions: Giotto, Rosetta, and PLATO

Europe's first deep space mission, Giotto, was launched in July 1985 to study Comet P/Halley. In addition to visiting its target comet, Giotto visited Comet P/Grigg-Skjellerup. Officially, the Giotto mission was terminated on July 23, 1992. In Comet P/Halley, the mission discovered hydrogen ions, water (which accounted for about 80 % by volume of all of the material thrown

[15] SMART-1http://www.esa.int/spaceinimages/Images/2006/07/SMART-1_trajectory_up_to_impact.

Fig. 8.3 Philae lander of the Rosetta mission (Graphic courtesy of ESA.)

out by the comet), carbon monoxide, carbon dioxide, methane, ammonia, hydrocarbons, iron, sodium, two major classes of dust particles (one dominated by the light "CHON" elements—carbon, hydrogen, oxygen and nitrogen, and the other rich in mineral-forming elements—sodium, magnesium, silicon, iron and calcium), and a layer of organic (carbon-rich) material covering the comet's surface.[16]

The subsequent deep space Rosetta mission was launched in March 2004, to "rendezvous with Comet 67P/Churyumov-Gerasimenko" and to "study the nucleus of the comet and its environment … and land a probe on its surface." This Philae lander was the first direct contact with a comet (see Fig. 8.3). The Rosetta-Philae mission was able to discover that water vapor from Comet 67P/Churyumov-Gerasimenko is significantly different from the water vapor found on Earth. The other findings were that the dust particles were released first when the comet was "fluffy" and that the asteroid contained a lot of sodium but no ice.[17]

ESA has active plans for yet another deep space mission. This is the planned launch of the PLAnetary Transits and Oscillations of stars (PLATO) mission to detect and characterize "terrestrial exoplanets around bright solar-type stars, with emphasis on planets orbiting in the habitable zone." The currently targeted launch date is 2024.[18]

[16] Fact Sheet, Giotto, ESA http://sci.esa.int/giotto/47355-fact-sheet Also see: Online: Giotto, NASA http://nssdc.gsfc.nasa.gov/nmc/masterCatalog.do?sc=1985-056A.

[17] ESA, News, "Rosetta Watches Comet Shed its Dusty Coat" (26 January 2015), online: Rosetta, ESA http://www.esa.int/Our_Activities/Space_Science/Rosetta/Rosetta_watches_comet_shed_its_dusty_coat.

[18] Online: Summary, PLATO, ESA http://sci.esa.int/plato/42276-summary/.

Summary of European Initiatives

The European Space Agency has engaged in an active space research program to explore Venus, Mars, Mercury, the Moon, Jupiter and its moons and deep space. These exploration activities have been comprehensive in their scientific scope and have included physical, chemical, and environmental measurements. The various missions have provided a large amount of data as to the composition and relative density of these various component bodies of the inner Solar System that could be of benefit to any future space mining activities.

Canada and Other National Initiatives

A number of other countries might have a clear or at least expressed an interest in space mining. This interest could be manifested in a variety of ways in the future. These countries include Canada, which is discussed in some detail below, but also Australia, Brazil, the Republic of Korea and South Africa as well as a number of others. There are further a number of mining, robotics, remote sensing, and even commercial space transportation companies from these countries that are also of interest. These countries or relevant companies might not engage in a complete space mining missions themselves but might seek partnerships with those planning such initiatives. Planetary Resources and Deep Space Industries in the United States, for instance, have been active in seeking partners with special expertise in key technologies, and such international partnerships may ultimately prove productive.

In an effort that was quite similar to Canada's NEOSSat mission, the German Space Agency (DLR) completed preliminary reviews for its DLR-AsteroidFinder mission, but this mission was canceled in 2012 due to excessive costs.[19] South Korea is participating in Japan's SELENE-2 mission and is planning to launch a lunar orbiter and lander in 2023 and a sample return mission in 2030.[20] A Netherlands-based not-for-profit company, Mars One, is considering a mission to transport the first four humans to the Red Planet by 2023 at a cost about $6 billion. A Phoenix-based not-for-profit organization, Astronauts4Hire, is providing training to its members as commercial "professional astronaut candidates who can assist researchers, payload developers, and spaceflight providers with mission planning and operations support."

As human history demonstrates, other countries and private sector entities—some serious and others not so serious—will try to follow these developments in an effort to search for natural resources, wealth, and colonization in outer space and on celestial bodies at least in our Solar System.

There are a number of countries that are developing commercial capabilities in robotics, tele-robotics and space transportation systems. Some of the

[19] http://www.dlr.de/fa/Portaldata/17/Resources/dokumente/abt_17/projekte/Handout_Asteroid_Finder.pdf (Last accessed: March 30, 2013).

[20] Srinivas Laxman, "Japan SELENE-2 Lunar Mission Planned For 2017, July 16, 2012, online: http://www.asianscientist.com/topnews/japan-announces-selene-2-lunar-mission-2017/ (Accessed last: September 1, 2015).

countries that are developing capability include the Republic of Korea, Israel, Ukraine, and Iran. A shortage of rare earth metals, the appeal of a new space industry, and the future importance of the technology associated with artificial intelligence, smart robotics systems, tele-operation, and remote power systems could all serve as incentives to move ahead with space mining initiatives around the world.

Canada is an example of a country that has shown considerable interest in space mining and is thus discussed in some detail below to illustrate a prime example of a country—plus its industry and academic and research community—that may, through partnerships with others become a player in future space mining operations.

The current strategy of the Canadian Space Agency (CSA) is to better understand the Solar System and the universe; to seek signs of life in extraterrestrial habitats, and to prepare for a permanent human presence in space and on other planets. In this regard, Canada is particularly interested in research focusing on the Martian atmosphere, terrestrial analogs and the search for life on Mars. Since Canada has extensive experience in robotics, its goal is to maintain and further develop such capabilities, including advanced mobility and in-orbit servicing systems. Canada is participating in NASA's Mars Sample Laboratory mission as well as the European/Russian ExoMars mission and the U. S. asteroid sample return mission, OSIRIS-Rex.[21] Within its robotic exploration program, Canada is planning to develop drilling and excavating capabilities to extract subsurface samples and conduct in-situ study of resources. In order to prepare for future exploration missions, dedicated sites in Canada where simulated conditions are analogous to those existing on the Moon and Mars are used. Canadian laboratories are testing technologies (e.g., drilling, rover navigation), and scientists are simulating water search, water-related land forming and search for life at those sites.

Canada's latest space exploration mission, the Near-Earth Object Surveillance Satellite (NEOSSat), was launched on February 25, 2013, aboard an Indian Polar Satellite Launch Vehicle (PSLV). This project was jointly funded by CSA and the Defense Research and Development Canada. It is designed to monitor asteroids as well orbiting space objects (including space debris) to help avoid collisions in space. The NEOSSat will help in cataloging near-Earth asteroids, thus "producing information that will be crucial to targeting new destinations for future space exploration missions." Eventually this information could prove quite helpful in making decisions about possible candidate asteroids for mining. [22]

CSA is believed to have spent nearly $110 million for the development of advanced robotics and space exploration technologies. As a result, CSA is introducing several rover prototypes that could be used for exploration missions

[21] See the OSIRIS-Rex mission website online: http://osiris-rex.lpl.arizona.edu/ (Last accessed: August 30, 2015).

[22] Miriam Kramer, "Suitcase-Size Satellite Launching Monday to Hunt Asteroids", February 24, 2013, http://www.space.com/19930-asteroid-tracking-satellite-neossat-launch.html (last accessed: 13 March 2013).

on the Moon or Mars.[23] The Minister of Industry, Christian Paradis, is reported to have said that "Canada's reputation for excellence has been carved out through decades of innovation and technological advances such as the iconic Canadarm, Canadarm2 and Dextre … . That legacy continues with the Next Generation Canadarm and these pioneer terrestrial rovers."[24]

Undoubtedly, Canada has been, and continues to be, very successful in niche areas of space activity (e.g., remote sensing, robotics). However, CSA faces perennial difficulty with limited funding and frequent budget cuts.[25] For example, CSA's budget for 2012–2013 is $388.3 million, but for 2013–2014 it will go down to $309.7 million, and for 2014–2015 it will only be $289.1 million.[26] This indicates that the Canadian government is not according high priority to, and investing sufficiently in, its space programs. This is indeed strange, particularly when viewed in the context of the regular increases in space budgets in countries such as China and India. Budgetary constraints limit the ability of the CSA to embark upon major initiatives on its own in the exploration and exploitation of space natural resources. In addition to the Canadian Space Agency, several institutions and companies are active in several aspects of the exploration of natural resources in space.

Canadian academic establishments are actively engaged in the exploration of these natural resources. The University of Western Ontario (UWO) is the leading institution in this regard. It hosts the Centre for Planetary Science & Exploration (CPSX) and is actively and extensively involved in education, outreach and research in planetary sciences, exploration and space system design using multidisciplinary perspectives. Its research themes include astrobiology, cosmochemistry, planetary atmospheres, planetary dynamics, planetary interiors, planetary surfaces, space systems, space robotics, tele-robotics, space history, exoplanets, and the monitoring of Earth-space. CPSX researchers are involved in NEOSSat and Mars Science Laboratory space missions.

The UWO is also home to the Canadian Lunar Research Network (CLRN). CLRN is a group of Canadian scientists, engineers, and entrepreneurs working collectively in order to promote lunar research and foster collaboration among Canadian and international researchers. Other members of the network are: Laurentian University, McGill University, Memorial University, the University of British Columbia, University of Guelph, University of New Brunswick, University of Toronto, the University of Winnipeg, York

[23] Megan Garber, "Curiosity's Cousins: Meet the Rover Fleet of the Canadian Space Agency – A fleet of vehicles ready to explore lunar and Martian terrains", OCT 23 2012, online: http://www.theatlantic.com/technology/archive/2012/10/curiositys-cousins-meet-the-rover-fleet-of-the-canadian-space-agency/263965/ (Last accessed: March 13, 2103).

[24] Merryl Azriel, "Canadian Space Agency Unveils Rover Fleet," October 22, 2012, online: http://www.spacesafetymagazine.com/2012/10/22/canadian-space-agency-unveils-rover-fleet/ (Accessed: March 13, 2013).

[25] The Canadian Press, "Canadian space programs face budget cuts, layoffs", Apr 9, 2012, online: http://www.cbc.ca/news/technology/story/2012/04/09/technology-csa-budget-cuts.html (Last accessed: March 13, 2013).

[26] CSA, 2012-2013 Report on Plans and Priorities, http://www.asc-csa.gc.ca/eng/publications/rpp-2012.asp (Accessed: March 13, 2013).

University, MDA, Optech, Odyssey MOON, NASA Lunar Science Institute, and OpenLuna. According to the website of the CLRN, the "current guiding research theme at the CLRN is lunar impacts—processes, products, and utilization. Impact cratering is considered to be the most important surface process on the Moon. As a result, research at CLRN focuses on understanding all aspects of lunar impact craters … . Looking both into the past and to the future, impact craters have been, and will be, key targets for the robotic and human exploration of the Moon."[27]

The Northern Centre for Advanced Technology (NORCAT) is a not-for-profit, non-share incorporated company located in Sudbury, Ontario, Canada. According to Chuck Black, NORCAT provides specialized mine training, occupational health and safety services and develops mining technology for space missions.[28]

Several Canadian professionals, think-tanks, mining and space industry associations, and business lobbying-groups are becoming quite active in deliberations concerning space and natural resources. What follows is a very brief description of such individuals and institutions and some of their recent activities.

The Canadian Institute of Mining, Metallurgy and Petroleum (CIM), a Westmount (Quebec)-based think tank, is organizing a convention under the theme of "Global Leadership—the Courage to Change" in Toronto from May 5–8, 2013. One of its six technical programs will concentrate on planetary mining sciences, in which four presentations are scheduled to be delivered, namely: "Helium 3 mining start-up transportation," by Thomas Taylor of Lunar Transportation Systems, Inc.; "Asteroid mining methods and opportunities for serving in space and terrestrial markets," by David Grump of Deep Space Industries; "Mining of PGMs in commercial space project management," by Christine Hansen of Hawaii Pacific University; and "Mining asteroid materials for manufacturing in low Earth orbit (LEO) and for valuable materials," by Michael Buet of Stott Space, Inc. The CIM, in collaboration with a team of five Canadian university professors, attended the 23rd World Mining Congress held in Montreal in August 2013.[29]

The Canadian Space Commerce Association (CSCA), a non-profit industry organization, is making efforts to advance the economic, legal and political environment for Canadian space-related companies. On March 7, 2013, CSCA ran a conference in Toronto under the theme Commercial Space Resource Utilization. The conference featured several interesting presentations related to technical, entrepreneurial, financial, regulatory and policy aspects of space natural resources exploration.[30]

---

[27] http://clrn.uwo.ca/research.htm (Accessed: August 27, 2015).

[28] E-mail from Charles Black of the Canadian Space Research Association. (April 2, 2013).

[29] The Canadian Institute of Mining, Metallurgy and Petroleum (CIM) "is the leading technical society of professionals in the Canadian Minerals, Metals, Materials and Energy Industries." Founded in 1898, the CIM aims "to facilitate the exchange of knowledge and technology [and] to foster networking, professional development and fraternity" in the minerals industry. http://www.cim.org/en/About-CIM.aspx (Accessed: April 2, 2013).

[30] Most of the presentations may be downloaded from: http://spacecommerce.ca/events/2013-csca-national-conference (Accessed: April 3, 2013).

There are several Canadian private companies that are involved in space-related activities, including several aspects of natural resources in space. Canada's largest aerospace company, McDonald, Dettwiler and Associates Ltd. (MDA), is vigorously involved in designing, building and supplying critical components and subsystems for several planetary mining exploratory missions from various countries. In addition, this highly experienced company is the prime contractor for the Odyssey Moon mission, which is "developing a sustainable commercial transportation system to deliver payload services to the Moon in support of science, exploration and commerce."[31] MDA supplied the advanced Instrument Deployment Device that is used on the Mars exploration rover mission, whose prime objective was to investigate the existence of water on Mars. "MDA's heritage in autonomous, safety-critical robotic systems for human and unmanned spaceflight is directly applied to rover developments across the range of exploration classes, from small, science-class autonomous prospecting rovers to large, tele-operable utility-class and manned vehicles. MDA has supported CSA, ESA, and NASA on over twenty Moon and Mars rover-related missions and system concept studies, technology developments, and analog field deployments."[32] Finally John Chapman, a private consultant and owner of JA Chapman Mining, has been enthusiastically active in this area.[33]

The Canadian Space Agency, experts, private sector, academic institutions and professional associations are increasingly becoming active in the idea of exploration of natural resources in space, although nobody has taken any concrete steps or embarked upon any serious and specific projects dedicated to this purpose similar to those announced by some private sector entities in the United States. Notwithstanding the foregoing, Canada seems to be ready to embark upon such exploratory activities once sufficient financial resources are available. Experts have expressed optimistic views that the exploration of natural resources in space will occur in the near future, as the potential for generating immense profits from such activities is becoming clearer and realistic. However, within Canadian commercial circles, there is a general apprehension about the current legal regime governing outer space. It is often stated that the inability to own the land and resources in outer space is the most serious barrier for private enterprises to be fully engaged in planetary exploration and exploitation of natural resources in space. It has also been suggested that since there is no mechanism for enforcement of international laws that govern the exploration and exploitation of these natural resources, those laws can be ignored.

[31] http://www.odysseymoon.com/ (Accessed: April 2, 2013).

[32] http://is.mdacorporation.com/mdais_canada/Offerings/Offerings_Rovers.aspx (Accessed: April 2, 2013).

[33] For more about Chapman's activities, see Charles Black, "Opportunities for Mining Operations on the Moon & Mars," http://acuriousguy.blogspot.ca/2012/02/opportunities-for-mining-operations-on.html?q=John+Chapman (Accessed: August 24, 2015).

It is believed that, with its extensive skills, expertise, technology and experience in terrestrial mining, Canadian industry should be able to take the lead in space mining. Although such credentials are important, one must keep in mind that the physical conditions in space (lack of gravity and atmosphere) are quite different from those that persist on Earth. Secondly, the legal status of outer space, asteroids and planets is significantly different from the legal status of Canadian territory, where most of such mining activities have occurred. These divergences pose serious challenges to, and must be taken into consideration by, the Canadian mining industry should it wish to enter the realm of space's natural resources.

Conclusions

It is still quite early days in the development of the field of space mining. There are significant legal, regulatory, and standards issues to overcome on top of the many technological, economic and business challenges that must be faced and overcome, before space mining becomes a reality. Nevertheless, Europe, plus Canada and others such as Australia, South Africa, the Republic of Korea, the Ukraine, Israel, and relevant industries from these countries could become significant players in the field over time. In short there is significant technological knowhow at the governmental and industrial level in these countries that could well be an important asset in the future.

9 Asian Space Programs: Japan, China and India

Japan, China and India are all assuming increasingly important roles in space as in many other high tech areas. As such, exploration, exploitation and utilization of space's natural resources are no longer left to the exclusive domain of the United States, the U.S.S.R./Russian Federation or the efforts of the European Space Agency and Europe's various national space agencies. The expanding capabilities around the world in the areas of human and robotic space exploration, advanced launcher systems and sophisticated planetary probes have created a new international context. In particular, this means that the exploitation of natural resources in space is no longer confined to the traditional science and technology domains and is now beginning to involve business and economic interests and even national competition. In short it is the issue of the future. And the future has become a truly international political issue. As a consequence, topics dealing with common concerns—sharing of resources, environment protection, and pollution—raise new difficulties, as they require different States to agree on common principles having a direct impact on their sovereignty and financial interests.

For many years the Asian national space programs were dominated by the activities of Japan that included those of the National Air and Space Development Agency (NASDA), the National Aerospace Labs, and broad range of planetary research programs carried out by Tokyo University's Institute of Space and Astronautical Science (ISAS). These wide-ranging activities of Japan were merged together as of October 1, 2003, to form a single integrated new Japanese agency known as JAXA—the Japan Aerospace eXploration Agency. But today there are rapidly growing and now very significant space programs in China and India as well. The Chinese program (the Chinese National Space Agency) and the Indian program (the Indian Space Research Organization) in their initial stages placed a great deal of emphasis on primarily developing launch vehicle technology. In these earlier days spacecraft systems were often provided by overseas suppliers. In the last 20 years all three of these Asian space programs have blossomed to

© Springer International Publishing Switzerland 2017
R.S. Jakhu et al., *Space Mining and Its Regulation*,
Astronautical Engineering, DOI 10.1007/978-3-319-39246-2_9

include the ability to design, engineer and oversee the domestic manufacture of spacecraft for exploration, science missions and applications. This chapter reviews the space exploration programs of these countries that now are supported by a range of proven launch vehicles and sophisticated space probe capabilities.

Japanese Space Exploration and Scientific Missions

The Institute of Space and Astronautical Science (ISAS) carried out many of Japan's early planetary research missions. Table 9.1 below provides a selected listing of ISAS planetary and solar probes that were built in the 1970s and 1980s. The particular highlight of ISAS came in 1990 when its MUSES-A probe successfully reached lunar orbit and collected data concerning the lunar surface (see Fig. 9.1).

An even more ambitious ISAS project was the Planet B or Nozumi probe that was launched in 1998 to undertake observations of the Martian atmosphere. The probe instrumentation failed, however, and thus the mission failed.

The last independent mission of ISAS before it merged to form JAXA

Table 9.1 Missions by ISAS from 1971 to 2003 prior to merging into JAXA

Launch date	Name before launch	Name after launch	Mission
Sept. 28, 1971	MS-F2	Shinsei	Ionosphere/cosmic-ray/solar-radio observation
Aug. 19, 1972	REXS	Denpa	Ionosphere/magnetosphere observation
Feb. 16, 1974	MS-T2	Tansei 2	Technology experiment
Feb. 24, 1975	SRATS	Taiyo	Thermosphere and sun
Feb. 4, 1978	EXOS-A	Kyokko	Aurora and ionosphere
Sept. 16, 1978	EXOS-B	Jikiken	Magnetosphere and thermosphere observation
Feb. 21, 1979	CORSA-b	Hakucho	X-ray astronomy
Feb. 21, 1981	ASTRO-A	Hinotori	Solar X-ray observation
Feb. 20, 1983	ASTRO-B	Tenma	X-ray astronomy
Jan. 8, 1985	MS-T5	Sakigake	Technology experiment/comet observation
Aug. 19, 1985	PLANET-A	SuiseiS	Comet observation
Aug. 19, 1987	ASTRO-C	Ginga	X-ray astronomy
Jan. 24, 1990	MUSES-A	Hiten	Interplanetary technology experiment
Aug. 30, 1991	SOLAR-A	Yohkoh	Solar X-ray observation (with NASA/UK)
July 24, 1992	GEOTAIL	GEOTAIL	Magnetosphere observation (with NASA)
Feb. 20, 1993	ASTRO-D	ASCA	X-ray astronomy (with NASA)
March18, 1995	SFU	SFU	Multi-purpose experiment flyer (with NASDA/NEDO and USEF)
Feb. 12, 1997	MUSES-B	Halca	Space VLBI technology development
July 4, 1998	PLANET-B	Nozumi	Mars atmosphere observation
May 9, 2003	MUSES-C	Hayabusa-1	Asteroid sample return technology development

History, Institute of Space and Astronautical Science https://en.wikipedia.org/wiki/Institute_of_Space_and_Astronautical_Science (Last accessed on August 30, 2015)

Fig. 9.1 The MUSES—a satellite designed by ISAS that Japan sent into lunar orbit (Graphic courtesy of JAXA.)

Table 9.2 Planetary research missions by ISAS after merging with JAXA

Launch date	Name before launch	Name after launch	Mission
July 10, 2005	ASTRO-EII	Susaku	X-ray astronomy
Aug. 24, 2005	INDEX	Reimei	Technology/Aurora research
Feb. 21, 2006	ASTRO-F	Akari	Infrared astronomy
Sept. 22, 2006	SOLAR-B	Hinode	Solar observation
Sept. 14, 2007	Selene	Kaguya	Lunar orbiter
May 20, 2010	PLANET-C	Akatsuki	Venus atmosphere observation
Sept. 14, 2013	SPRINT-A	Hisaki	EUV observation
Expected in 2016	Astro-H		X-ray astronomy
Expected in 2016	ERG		Magnetosphere research
Expected in 2016	MMO	MMO	Mercury exploration (part of BepiColombo, with ESA)

History, Institute of Space and Astronautical Science https://en.wikipedia.org/wiki/Institute_of_Space_and_Astronautical_Science (Last accessed: August 30, 2015)

was MUSES-C, a technology development project intended to attempt a planetary sample return. As an independent research institute, ISAS also developed the M-V rocket services that launched a number of their planetary missions, including the *Hayabusa 1*.[1]

Even after the formation of JAXA, ISAS continues to play the leading role in Japan's planetary research programs. The important planetary-related probes by ISAS since 2003 includes those missions provided in Table 9.2.

In 2005 JAXA, the Japanese national space agency, shortly after its official formation, developed its own vision known as "Vision 2025." The JAXA

[1] M-V 1-8 Japanese Rocket System, http://historicspacecraft.com/Rockets_Japanese.html.

vision establishes the goals for the 20-year period spanning 2005–2025. Vision 2025 details JAXA's longer terms goals in which space exploration plays a major role. Under the Vision, Japan aims to turn itself into a "top science center" for space science; to develop sound technologies for the establishment and utilization of a lunar base; and, to establish a "deep space harbor" on the Moon and/or at a Lagrange point in the future.

Under the Vision 2025 plan Japan indicated its goals of:

- Promoting studies for possible utilization of the Moon.
- Developing cutting-edge technologies such as robotic technologies, nanotechnologies and micro-machines.
- Researching and developing power-providing technologies using solar power.
- Preparing for the establishment of a human lunar base.
- Developing complementary relationships with other nations for effective exploration of space.

The vision envisages that further exploration of the Moon with Moon-orbiting satellites will take place within the first 10 years. In this timeframe, Japan will also seek a decision by the government on whether to take significant steps toward the utilization of the Moon. After 20 years, the Vision anticipates that Japan will contribute to the international community by taking up roles in the implementation of international lunar initiatives.[2]

[2] JAXA2025 Plan, http://gidicoded.com/site_video-download.xhtml?get-id=jgs8G_EScz4&title=JAXA-2025-JAXA-Longterm-Vision.

JAXA's Selenological and Engineering Explorer mission (SELENE/Kaguya), launched in September 2007—just before the Chinese and Indian lunar missions were launched. Its mission was to understand the Moon's origin and evolution. SELENE/Kaguya carried 14 different scientific instruments that operated quite successfully until June 2009.[3] The mission is to further investigate the Moon to obtain information on its composition, geography, surface and sub-surface structure.[4] It will also help to define whether the use of the Moon's natural resources is possible in the future. JAXA plans for SELENE to be the largest lunar mission since the Apollo program.

JAXA also has plans for SELENE-2—a mission that will consist of an orbiter, a lander and a rover that will conduct in-situ studies on the Moon. SELENE-2 was originally slated for launch in 2017 but will now likely happen in 2018 or later. According to Tatsuaki Okada of JAXA, SELENE-2 is designed "for developing … key technologies for future human exploration. It is a multipurpose mission which is a precursor for human exploration."[5]

SELENE-2, or the Selenological and Engineering Explorer 2, is a planned

[3] Viorel Badescu, Foreword, Moon: prospective energy and material resources, Springer, online: http://link.springer.com/book/10.1007/978-3-642-27969-0/page/1 (Last accessed: March 13, 2013).

[4] Selene project, JAXA www.jaxa.jp/missions/projects/sat/exploration/selene/index_e.html (Last accessed: August 29, 2015).

[5] Srinivas Laxman, "Japan SELENE-2 Lunar Mission Planned For 2017", July 16, 2012, online: http://www.asianscientist.com/topnews/japan-announces-selene-2-lunar-mission-2017/ (Last accessed: March 13, 2013).

Fig. 9.2 SELENE-2 shown lowering to lunar surface (Graphic courtesy of JAXA.)

Japanese robotic lunar spacecraft that will include an orbiter, a lander and a rover. It is expected to be launched in 2017 as a successor to the 2007 SELENE (Kaguya) lunar orbiter. This is a quite expensive but also prestigious program that is considered key to Japanese future plans to establish a lunar colony and perhaps to use robotic systems for engaging in space mining in the future (see Fig. 9.2).

While the legal and regulatory issues about whether and how planetary resources might obtained and returned to Earth are settled, the technological development keeps moving ahead. This is not only on Japan's future agenda, but as will be addressed later in this chapter China and India are also pursuing relevant space system capabilities as well.

Japan has also been involved in asteroid exploration with its $150 million *Hayabusa-1* mission—the first asteroid mineral prospecting spacecraft that successfully completed its mission in June 2010 and returned to Earth with a sample.[6] In January 2012, JAXA made its first international announcement of opportunity relating to the study of the samples,[7] and a second announcement of opportunity was issued on January 9, 2013.[8]

As a follow-up to *Hayabusa-1*, JAXA has planned a new $400 million asteroid

[6] JAXA. "Hayabusa." http://www.jaxa.jp/projects/sat/muses_c/index_e.html http://www.hayabusa.isas.jaxa.jp/e/index.html (Last assessed: August 30, 2015).

[7] JAXA Press Release: 1st International Announcement of Opportunity for Hayabusa sample investigation (24 January 2012). JAXA website online: http://www.jaxa.jp/press/2012/01/20120124_hayabusa_e.html (Last accessed: March 13, 2013).

[8] http://hayabusaao.isas.jaxa.jp/ (Last accessed: August 30, 2015).

Fig. 9.3 Artist's concept of *Hayabusa-2* probe shown taking samples of Asteroid 1999 JU3 in mid-2018 (Graphic courtesy of JAXA.)

mission, *Hayabusa-2*. JAXA anticipates that the spacecraft will arrive at the carbonaceous asteroid identified as 1999 JU3 in mid-2018, and that it will likely return to Earth at the end of 2020. Asteroid 1999 JU3, a 920-m (3010-ft) large space rock, was chosen "because it is widely believed to contain organic matter, such as that which may have contributed to life on Earth"[9] (see Fig. 9.3).

Finally, as noted in the earlier chapter on European missions, Japan is a key participant along with the ESA in the Mercury exploration mission known as BepiColombo, which is planned to be launched in 2016. This mission will conduct comprehensive observations of the planet Mercury that will help to determine how much it has in common with other planets and what elements are unique to Mercury as well as the origin and evolution of other planets.[10]

Japan clearly has strategic longer-term objectives in space. This has been shown in the active space missions carried out by the Institute of Space and Astronautical Science (ISAS) since the

[9]"JAXA Schedules New Asteroid Sampling Mission for 2014", online: http://www.spacesafetymagazine.com/2013/01/03/jaxa-schedules-asteroid-sampling-mission-2014/ (last accessed: 15 March 2015). http://www.isas.jaxa.jp/e/index.shtml (Last accessed: August 30, 2015).

[10]"Bepicolombo Mercury Mission to be Launched in 2015", 28 February 2012, online: http://sci.esa.int/science-e/www/object/index.cfm?fobjectid=50105 (Last accessed: August 28, 2015).

early 1970s, the ambitious JAXA 2025 strategic plan to develop space robotic capabilities and explore the Moon, and ultimately to create a permanent space presence there. Japan's ambitious SELENE-1 and 2 missions, plus its significant asteroid research program, as reflected in *Hayabusa 1* and *2*, all indicate that resource-challenged Japan sees the possibility of space mining as a strategic goal to be pursued by an active space research and exploration program.

Japan consolidated its launcher development program with the creation of JAXA and with discontinuing the M-V launcher program of ISAS. This led to the development of the H-1, H-II, H-IIA and HII-B program as well as the H-transfer vehicle. There were with two failures with the H-II launcher in 1998 and 1999. This led to a Japanese Diet-mandated international review committee that led to a series of changes in the quality assurance program, which eventually produced excellent results. In the last 15 years there has been increasing success with larger and higher capacity H-II A and H-IIB launchers and with successful trips of the HTV to the International Space Station with resupply missions.[11]

[11] Space Launch Vehicle Reliability http://www.ewp.rpi.edu/hartford/users/papers/engr/ernesto/cedenc/SMRE/Project/Space%20Shuttle%20Vehicle%20Reliability.pdf The 17 member international launch reliability review committee with 9 Japanese and 8 international members from Europe and the US recommended that the quality assurance and reliability review processes be separated from the line management so that reliability concerns could be independently reported.

China's Planetary Research and Exploration Programs

China also commenced a space program in the 1970s. In line with the needs of the country, China developed a fleet of satellites for space applications (often using international satellite procurement to accomplish this task) and placed the greatest emphasis on perfecting reliability and performance of the CZ and Long March family of launchers. Based on its close relationship with Russia, China obtained considerable technical assistance from Russia to develop both its space infrastructure and technology, including its now quite successful human spaceflight program. In 1976 China became a member of Intelsat and also became more involved with Western space programs as well. In the 1990s China significantly increased the level of activity of its national space program and shifted away from placing its greatest emphasis on space applications to developing a range of technologies to support a human space exploration program.

The successful launch of a Chinese taikonaut on October 15, 2003, aboard a Long March F3, marked a major milestone in the development of China's space program. It became the third country in the world after the U.S.S.R./Russia and the United States to successfully launch astronauts into space.[12]

China plans to conduct deep-space exploration of the Moon in a series of stages. According to its 2011 White

[12] Making History: China's First Human Spaceflight, Space.com, September 28, 2005 http://www.space.com/1616-making-history-china-human-spaceflight.html (Accessed: August 30, 2015).

Paper on China's Space Activities this will involve first orbiting, then landing, and finally exploring and returning samples. This strategic space planning document states that "China's lunar probe projects are based on .... orbiting, landing and returning .... In the third stage, China will start to sample the Moon's surface ... and return those samples back to Earth." Natural resources from space will play a key role of this exploration strategy. According to Ouyang Ziyuan, chief scientist of China's lunar exploration program, the Moon "probably holds the key to humanity's future subsistence and development." Luan Enjie, director of China's National Space Administration, has also mentioned that China would be interested in exploiting rare resources found on the Moon's surface.[13]

China launched *Chang'e-1* in 2007, at a cost of 1.4 billion yuan (equivalent to approximately U. S. $170 million), to provide a map of the Moon and also to evaluate helium-3 resources on the Moon. Launched in 2010, *Chang'e-2* carried more sophisticated instruments on board. According to the 2011 White Paper on China's Space Activities: "*Chang'e-2,* created a full higher-resolution map of the Moon, and a high-definition image of Sinus Iridium, and completed several extended tests, including circling the Lagrangian Point L2, which laid the foundation for future deep-space exploration tasks."[14]

China's three-phase Moon exploration program is both thorough and ambitious. China's *Chang'e-3* with its Yutu Moon rover was launched in December 2013. This Moon rover was equipped with "lunar exploration radar" as well as a drilling machine to excavate Moon rocks from a depth of 2 m below the surface of the Moon. Despite some initial problems it has remained operational for over a year[15] (see Fig. 9.4).

China's Moon exploration program includes plans "to launch ... its fifth lunar probe, *Chang'e-5* in 2017 while *Chang'e-6* might be launched in 2018.[16] These two missions are both being equipped so to send back a Moon rock sample to Earth."

China's first Mars mission, *Yinghuo-1,* was a joint venture in cooperation with the Russian Phobos-Grunt mission. This spacecraft was designed to probe the environmental and magnetic fields of Mars. With the failure of the Phobos-Grunt, China now plans to wait until 2016 to launch a Mars probe by itself.[17]

---

[13] "China to launch orbiters for lunar soft landing in next five years: white paper", 29 December 2011, http://news.xinhuanet.com/english/china/2011-12/29/c_131333253.htm (Last accessed: August 30, 2015).

[14] NASA.gov http://nssdc.gsfc.nasa.gov/nmc/spacecraftDisplay.do?id=2007-051A (Accessed last: August 15, 2015); "China plans moon landing around 2017" China News (November 5 2005), http://news.xinhuanet.com/english/2005-11/05/content_3733767.htm (Last accessed: August 15, 2015).

[15] Spaceflight 101, March 9, 2015, "China's Yutu Rover still responsive after 64 Weeks on the Moon" http://www.spaceflight101.com/change-3-mission-updates.html (Last accessed: August 30, 2015).

[16] "China Expects to Launch Fifth Lunar Probe Change-5 In 2017," Mar 3, 2011, http://www.moondaily.com/reports/China_Expects_To_Launch_Fifth_Lunar_Probe_Change5_In_2017_999.html (Last accessed: August 30, 2015).

[17] "China's 1st interplanetary probe hits mars mission", January 20, 2012, http://zeenews.india.com/news/space/china-s-1st-interplanetary-probe-hits-mars-mission_753748.html

Fig. 9.4 The Yutu Moon rover from the *Chang'e-3* mission on the surface of the Moon (Graphic courtesy of JAXA.)

China's Moon and Mars exploration activities constitute an important part of worldwide efforts for the exploration of space's natural resources. In technological terms, China is still behind the United States, yet is constantly accelerating its space capabilities.

However, due to its booming economy; geo-political "competition" with the United States, India and Japan; and the insatiable global need for natural resources, China seems determined to make a speedy and steady progress towards becoming a significant player and an active voice in international space discussions, including those related to the exploration of natural resources in space. The one constant in the Chinese space program is that it has always been quite pragmatic in concentrating first and foremost on practical applications—first on telecommunications, broadcasting, remote sensing, space navigation, and meteorological satellites and now with a prime focus on the Moon that might be the future source of rare earth metals and energy source materials such as helium-3.

The Indian Space Program

In recent years, India's space program has undergone a major change in focus from original developmental space applications to the pursuit of space science and exploratory programs, including ambitious space missions. In this

(accessed last: August 30, 2015); "Will China send its own probe to Mars?," August 7, 2012, http://english.cntv.cn/program/newsupdate/20120807/100268.shtml (Last accessed: August 30, 2015).

regard it is a parallel to both the Indian and the Chinese space programs.

India is planning to launch scientific missions to space on a regular basis as well as develop a human spaceflight program in the not-too-distant future. The Indian government's consistent support for space activities has been partially induced by its efforts to increase India's stature in the world and the largely unacknowledged competition with its northern neighbor, China. According to James Clay Moltz of the U. S. Naval Postgraduate School's Department of National Security Affairs in Monterey, California, there is "definitely a space race between India and China for top regional prestige and influence … . Evidence shows the two countries are watching each other's activities in space very carefully and keeping score as to who is gaining and losing influence."[18]

From its humble beginnings in 1969, the Indian Space Research Organization (ISRO) has achieved an enviable status as the world's sixth top space organization. "While its first fifty missions took twenty-seven years, the next fifty took place in the last ten years. Under current planning nearly 60 missions will happen within a concentrated span of five years." Analysts point out that though the ISRO's budget is less than one-tenth of NASA's, it has been increased every year since the early 2000s, jumping from the equivalent of U. S. $590 million in 2005, to about $1.5 billion currently.[19]

India's first scientific mission, *Chandrayaan-1,* was launched by ISRO in October 2008. This unmanned mission to the Moon was accomplished for only U. S. $70 million. This lunar mission was achieved for considerably less than similar missions launched by the Japanese and Chinese around the same time. Its objectives were to advance Indian technological capabilities and experience; prepare India for future uses of the Moon; and develop high resolution remote sensing images of the Moon, to map out the Moon in terms of elements, minerals and topography.[20]

India's first mission to Mars was accomplished on project budget of only U. S. $90 million. The Mars orbiter mission, successfully launched in October 2013, is equipped with five highly capable experimental payloads, including "the Methane Sensor, Thermal Infrared Spectrometer, Mars Color Camera, Lyman alpha photometer and Mars Exospheric Neutral composition analyzer." The crucial objective of the mission, of simply putting a satellite into Mars orbit, was brilliantly achieved. Prior to the mission, industry observers expressed the view that "simply to succeed at launching a spacecraft on an Earth-to-Mars transfer orbit, successfully navigate it to Mars, successfully

[18] Neeta Lal, "India's Space Ambitions, Space race picks up even as economy flags" 08 March 2013, ASIASENTINEL, http://www.asiasentinel.com/index.php?option=com_content&task=view&id=5238&Itemid=404 (Last accessed: August 30, 2015).

[19] Neeta Lal, "India's Space Ambitions, Space race picks up even as economy flags" 08 March 2013, ASIASENTINEL, http://www.asiasentinel.com/index.php?option=com_content&task=view&id=5238&Itemid=404 (Last accessed: August 30, 2015).

[20] India launches first Moon mission,"22 October 2008, http://news.bbc.co.uk/2/hi/science/nature/7679818.stm (Last accessed: August 30, 2015).

Fig. 9.5 Image of Mangala Valles region of Mars from Indian Mars orbital mission (Graphic courtesy of the Indian Space Research Organization, ISRO)

enter orbit at Mars, and operate it there at all would be major achievements for India, regardless of any scientific data return (Fig. 9.5)."[21]

The *Chandrayaan-2* mission for detailed exploration of the Moon has a projected budget of just under U. S. $80 million.[22] This mission was initially planned as a joint Indian-Russian venture. It was delayed primarily due to the December 2011 failure of the Russian Phobos-Grunt mission. Since Russia has pulled out from the joint venture, India is going ahead alone with a currently projected launch date of either late 2017 or early 2018.

Chandrayaan-2 (in Sanskrit *Chandrayaan* means "Moon-vehicle") is the follow on to *Chandrayaan-1*. This mission, in its current form, was entirely developed by the Indian Space Research Organization (ISRO). The mission is planned to be launched to the Moon by a geosynchronous satellite launch vehicle (GSLV). It includes a lunar orbiter, a lander and a lunar rover, all developed by India.

The objective of this mission will be to study specific sites such as the far side of the Moon. According to ISRO, the scientific goals of the mission are: "to further improve the understanding of the origin and evolution of the Moon using instruments onboard [the] Orbiter and in-situ analysis of lunar samples using [the] Lander and [the] Rover." Two of the five payloads selected for the orbiter are: "(1)

[21] "India's Mars Mission Current Status" Spaceflight101, May 30, 2015 http://www.spaceflight101.com/mars-orbiter-mission-updates.html (Last accessed: August 30, 2015).

[22] "Chandrayaan-2 to cost Rs 426 crore," 9 April 2011, online: http://www.indianexpress.com/news/chandrayaan2-to-cost-rs-426-crore/773971 (Last accessed: March 13, 2013).

Imaging IR Spectrometer (IIRS) for mapping of [the] lunar surface over a wide wavelength range for the study of minerals, water molecules and hydroxyl present; (2) Terrain Mapping Camera-2 (TMC-2) for preparing a three-dimensional map essential for studying the lunar mineralogy and geology."[23]

Finally, India is planning for its first manned mission with the objective of launching a crew of three Indian astronauts into low Earth orbit with a successful return to Earth.[24] The Indian crew module is temporarily named Orbital Vehicle. It is currently intended to be the basis of the several "manned missions." This space capsule (or "module") is designed to carry three people, and a planned upgraded version will be equipped with rendezvous and docking capability. In its maiden manned mission, this largely autonomous 3.7-ton capsule will orbit Earth at a 400-km (250-mile) altitude. The crew vehicle is planned to be launched on ISRO's geosynchronous launch vehicle Mark III. This Hindustan Aeronautics Limited (HAL)-manufactured crew module had its first unmanned experimental flight on December 18, 2014, and the flight was successful[25] (see Fig. 9.6).

[23] R. Ramachandran, "Chandrayaan-2: India to go it alone" January 22, 2013, online: http://www.thehindu.com/news/national/chandrayaan2-india-to-go-it-alone/article4329844.ece (Last accessed: August 30, 2015).

[24] Srinivas Laxman, "Japan SELENE-2 Lunar Mission Planned For 2017, July 16, 2012, online: http://www.asianscientist.com/topnews/japan-announces-selene-2-lunar-mission-2017/ (Last accessed: March 13, 2013).

[25] K.S. Jayaraman (11 February 2009), Designs for India's First Manned Spaceship Revealed, Bangalore: Space.com (Last accessed: August 30, 2005).

If these launches continue to be nominal and low Earth orbit manned flights continue as planned then there may be a subsequent plan to send a crew to the Moon. The status of such a mission is currently unclear, with no specific launch date.

As human history demonstrates, other countries and private sector entities—some serious and others not so serious—will try to follow these developments in an effort to search for natural resources, wealth, and colonization in outer space and on celestial bodies at least in our Solar System.

Conclusions

The foregoing discussion clearly demonstrates that in the national space policies and programs of the major spacefaring countries, there is a strong focus on space exploration but with some clear practical intent behind many of the missions that focus primarily on the Moon, Mars and asteroids. A major common objective can be found in the desire to learn about natural resources in space in order to determine their utility and relevance for purposes of future exploitation. Exploitation of such natural resources has not yet occurred, but its feasibility is becoming clearer as additional space launch technology, robotic mining systems, remote power systems and tele-operations are developed. When exploitation of these natural resources begins, it is critical that the major legal, regulatory and standards issues be resolved—hopefully on the basis of a global consensus.

It is assumed by many that the use of the natural resources in space and their exploitation will indeed materialize at

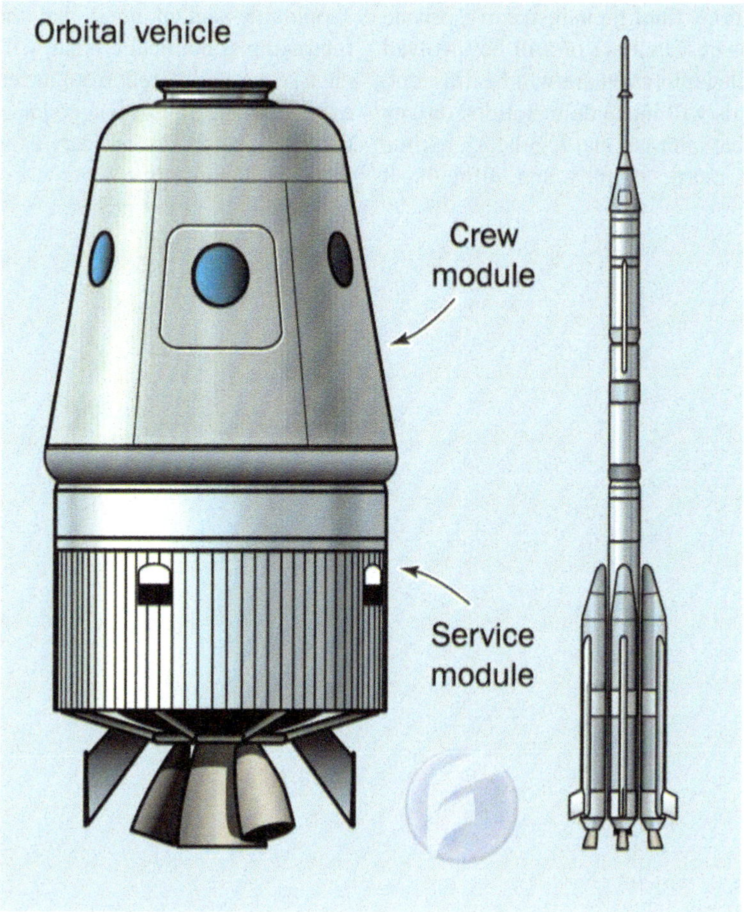

Fig. 9.6 Indian Orbital Vehicle and GSLV Mark III launch vehicle (Graphic courtesy of the Indian Space Research Organization, ISRO.)

some point—perhaps in the relatively near future. Linked to this hypothesis, it is also assumed that governments will remain key actors in space while the private sector will also continue to expand its role in space. The future potential exploitation and use of these natural resources will not likely be left in the hands of the private sector, even if they should ultimately play a significant role in space mining operations. This strongly suggests the need for a good and reasonable balance between the interests of private investors and those in the public sector with protection of the rights of all that are involved.

Rivalries and competition coupled with cooperation in various forms seems to be the order of the day, particularly in space exploration missions for natural resources. This certainly seems to be true as between China and the United States as well India and China. The preferences and interests of a diverse group of nations (and their respective private companies) that are or will be involved in such exploration are/will be different, and this will make deliberations on any possible international regulatory regime much more complex and difficult. It should be kept in mind that the new international political climate will present a scenario different from the one that existed at the time of negotiations for the 1967 Outer Space Treaty as well as the 1979 Moon Agreement.

10 The International Legal Framework

As described in detail in the preceding chapters, beyond exploration there is also a clear and sustained desire among spacefaring nations to prospect for and eventually exploit the natural resources found in space.[1] The emergence and increasing involvement of the private (i.e., non-governmental) sector in the conduct of space activities is also an important trend that is growing exponentially in scope. Exploration for these natural resources has become a truly international issue with far reaching political, commercial and strategic implications. As a consequence, common international concerns, such as the freedom of exploiting space for its natural resources, the sharing of benefits of such exploitation, orderly and sustainable exploitation, environmental protection, safety standards and procedures, and the difficulty in achieving consensus on the legal principles governing such activities, will undoubtedly have a direct impact on the varied political and commercial interests of States and other stakeholders.

There are several central issues here. First, there is the issue of whether or not outer space truly constitutes a "global commons." Secondly, there are many pending concerns about the nature of enforcement systems in space and whether this will ultimately lead to the deployment of weapons or defensive forces in outer space. Thirdly, there is the question of the future size and import of a space economy and whether this will ultimately be a significant part of world economic interests.

Although it has not yet occurred as of the time of this writing, the exploitation of natural resources in space is becoming increasingly feasible. When the exploitation of these natural resources begins in earnest, the legal issues and implications arising from them will be critical to its success. As such, it is important to proactively anticipate and

[1] The term '*space exploration*' refers to all activities related to discoveries in outer space and natural resources of the planets (other than those of the Earth) by the use of robotics and human space flights. The term '*exploitation*' on the other hand means extraction and refinement of natural resources essentially for commercial purposes. See: Ram Jakhu, *Twenty Years of the Moon Agreement: Space Law Challenges for Returning to the Moon* (2005) Zeitschrift für Luft- und Weltraumrecht, 244.

© Springer International Publishing Switzerland 2017

R.S. Jakhu et al., *Space Mining and Its Regulation*,
Astronautical Engineering, DOI 10.1007/978-3-319-39246-2_10

address some of those legal issues in order to ensure that the exploitation of these natural resources occurs in an orderly fashion when it actually becomes practicable. This chapter describes and analyzes the existing international legal framework (and the shortcomings thereof) that governs the exploration, use and exploitation of natural resources in space.

International space law governs the conduct of space activities of any kind, including the exploration, use and exploitation of the natural resources of outer space and of celestial bodies. Many of the principles, rules and regulations of international space law have been codified in the five space law treaties adopted under the auspices of the United Nation's Committee on Peaceful Uses of Outer Space (UN COPUOS) between 1967 and 1979.[2] The 1967 Outer Space Treaty is the first and the most widely accepted of the five U. N. space law agreements.[3] Its purpose was to establish general principles to be applied prospectively to govern space activity. It therefore creates binding legal obligations for the States that are party to it. Indeed, some of the principles contained in the 1967 Outer Space Treaty have become customary international law as they have been widely accepted by the international community, and State practice associated with those principles has also been consistent.[4] Such principles are therefore equally applicable to States' parties as well as non-States' parties to the 1967 Outer Space Treaty.

The 1979 Moon Agreement is the fifth and last international space law instrument adopted under the auspices of the UN COPUOS. It was specifically negotiated and adopted to set out principles and rules governing humankind's exploration, use and exploitation of the natural resources of the Moon and other celestial bodies. So far, the Moon Agreement has been signed by four States and ratified by 16.[5] It entered into force on July 11, 1984, but the scope of application of its provisions is limited

[2] These are: *Treaty on Principles Governing the Activities of States in the Exploration and Use of Outer Space, including the Moon and Other Celestial Bodies* 27 January 1967 18 UST 2410; TIAS 6347; 610 UNTS 205 [*1967 Outer Space Treaty*]; *Agreement on the Rescue of Astronauts, the Return of Astronauts and the Return of Objects Launched into Outer Space* 22 April 1968 19 UST 7570; TIAS 6599; 672 UNTS 119 [*1968 Rescue Agreement*]; *Convention on International Liability for Damage Caused by Space Objects* 29 March 1972 24 UST 2389; TIAS 7762; 961 UNTS 187 [*1972 Liability Convention*]; *Convention on Registration of Objects Launched into Outer Space* 14 January 1975 28 UST 695; TIAS 8480; 1023 UNTS 15 [1975 Registration Convention]; and, the *Agreement Governing the Activities of States on the Moon and Other Celestial Bodies* 18 December 1979 18 ILM 1434; 1363 UNTS 3 [*1979 Moon Agreement*].

[3] As of 4 April 2016, there were 104 States Parties to this Treaty and 25 other States that have signed but not ratified it. For the text of the Treaty, see Appendix to this book.

[4] See Ronald L. Spencer, Jr, "International Space Law: A Basis for National Regulation" in Ram S Jakhu, ed, *National Regulation of Space Activities* (Heidelberg: Springer, 2010), p. 1.

[5] As of 4 April 2016, there were 16 States Parties to the Moon Agreement. The 4 other States that have signed but not ratified it are: France, Guatemala, India and Romania. For the text of the Agreement, see Appendix to this book.

exclusively to the 16 States Parties presently. The 1967 Outer Space Treaty and the 1979 Moon Agreement are the two U. N. space law agreements that directly address the exploration, use and exploitation of the natural resources of the Moon and other celestial bodies. As such, this chapter only analyses the provisions of these two treaties.

For a proper understanding of the applicable law in its present form (*lex lata*), it is imperative to follow the general rule for the interpretation of treaties as set out in Article 31(1) of the Vienna Convention on the Law of Treaties.[6] The provision provides that "A treaty shall be interpreted in good faith in accordance with the ordinary meaning to be given to the terms of the treaty in their context and in the light of its object and purpose." Moreover, the law as it is (*lex lata*) must not be confused with the law as it should be (*lex de ferenda*) or the law as it should have been. In addition, it should be kept in mind that an international treaty is usually a result of compromises between the different views and perspectives held and advanced by the nations involved in the process of negotiating the treaty. These compromises are incorporated in various parts or articles of that treaty. Therefore, in order to fully understand the precise and accurate nature and scope of a treaty provision setting out a principle or rule of law, one must not focus exclusively on the language of a single provision or article but must read the provision or article in the context of, or in relation to, the other provisions or articles of the treaty. In his testimony before the U. S. Senate's Committee on Foreign Relations on the Outer Space Treaty, Ambassador Arthur Goldberg, who had participated in the treaty negotiations on behalf of the United States, correctly identified the above-mentioned principles and underlined that "any document must be read in its entirety, and you must take article I and read it in reference to articles II, III, IV, the whole Treaty. You cannot isolate one section and read it in isolation, and when you read it as a whole, you get the meaning of the Treaty."[7] In this chapter we follow this approach in order to provide a clear understanding of the precise nature and scope of the provisions of both the Outer Space Treaty and the Moon Agreement.

International space law constitutes a specialized branch of international law. As such, the default rule is that recourse may be had to general principles of international law to resolve novel situations that are not specifically addressed by the existing specialized body of international space law. In addition to the provisions of the five U. N. space law treaties, there are many other international legal principles, rules and increasingly, guidelines (or what has been termed by some commentators as "soft law" rules) that also govern the conduct of space activities at the international level. To the extent that they may be adapted to suit the unique environment of outer space, including the celestial bodies, some of the principles

[6] *Vienna Convention on the Law of Treaties*, 23 May 1969, 1155 UNTS 331 [*VCLT*].

[7] *Treaty on Outer Space,* Hearings Before The Committee on Foreign Relations, United States Senate, Ninetieth Congress, First Session, Executive, D, 90th Congress, First Session, March 7, 13 And April 12, 1967, pp. 33 and 34.

of general international law that govern resource exploitation and environmental protection in the terrestrial "global commons" could provide a basis for a future international governance regime for the exploitation of the natural resources found in space. In resorting to these general principles, however, extreme care must be taken to avoid transposing legal regimes that have been developed specifically to govern the conduct of terrestrial activities into space on a wholesale basis. Any such exercise is bound to fail, since the physical environment of outer space is unique and different from other global commons.

The 1967 Outer Space Treaty

In 1957, the space age ushered in a new and unprecedented era not only in global scientific and technological progress but also in international law. The international community acknowledged this fact in 1961 under a U. N. General Assembly resolution, which underscored the following principles:

(a) the common interest of mankind in furthering the peaceful use of outer space and the urgent need to strengthen international cooperation in this important field;
(b) the exploration and use of outer space should be exclusively carried out for the betterment of mankind and to the benefit of all countries, irrespective of the stage of their economic or scientific development; and,
(c) outer space and celestial bodies are free for exploration and use by all States in conformity with international law and are not subject to national appropriation.[8]

These stipulations were incorporated with minor refinements into the operative part of the 1963 resolution of the U. N. General Assembly,[9] as well as in the binding provisions of the Outer Space Treaty, thereby affording them the status of customary international law. The U. N. General Assembly unanimously adopted the Outer Space Treaty in 1967. The unanimous adoption of this treaty highlights the agreement between States on the need to achieve broad international cooperation in the scientific as well as the legal aspects of the exploration and use of outer space for peaceful purposes, as well as for the development of mutual understanding and the strengthening of friendly relations between States and peoples.[10]

The Common Interest Principle and Freedom of Exploration and Use of Outer Space

Article I of the Outer Space Treaty provides in relevant part as follows:

> *The exploration and use of outer space, including the Moon and other celestial bodies, shall be carried out for the benefit and in the interests of all countries, irrespective of their degree of economic*

[8] UN General Assembly Resolution 1721 (XVI) of *20 December 1961*.

[9] UN General Assembly, *Declaration of Legal Principles Governing the Activities of States in the Exploration and Use of Outer Space*, Resolution 1962 (XVIII), adopted without vote on 13 December 1963.

[10] *Outer Space Treaty*, Preamble.

> *or scientific development, and shall be the province of all mankind.*
>
> *Outer space, including the Moon and other celestial bodies, shall be free for exploration and use by all States without discrimination of any kind, on a basis of equality and in accordance with international law, and there shall be free access to all areas of celestial bodies.*

The first paragraph of Article I of the Outer Space Treaty contains the so-called common interest principle. It was proposed in the UN COPUOS by developing countries (led by Brazil) in order to legally protect and guarantee their future right to explore and use outer space and the celestial bodies when they eventually become economically and scientifically capable of doing so. It may be noted that the incorporation of this principle in the binding or operative part of the treaty, rather than in the non-binding or aspirational preamble, was the result of one of the compromises that paved way for the final adoption of the treaty. After the completion of the draft treaty in the UN COPUOS, the U. S. delegate stated that:

> *[the] spirit of compromise shown by the space Powers and the other Powers had produced a treaty which established a fair balance between the interests and obligations of all concerned, including the countries which had as yet undertaken no space activities ... [Article I para 1] like the provision prohibiting national appropriation by claim of sovereignty, was a strong safeguard for those States which at present had no space programme of their own.*[11]

Despite the recognized need to achieve a balance of interests, there are several definitional issues that require clarification. Firstly, there is no precise definition of the term "celestial bodies." Does it include all small and large bodies in space, or does it refer only to large planetary bodies and natural satellites, such as the Moon? In the absence of an internationally agreed upon definition, the ordinary meaning of "celestial bodies" should be used. According to *Universe Today*, "a celestial body is any natural body outside of the Earth's atmosphere."[12] Thus, any asteroid is a celestial body irrespective of whether it is known or unknown, its dimensions, orbit, speed and distance from Earth. The asteroid Ceres, with its size approximately equal to the state of Texas, is "a celestial body that is by far the largest and most massive asteroid in the belt between Mars and Jupiter .... The asteroid Cruithne is sort of small and indistinct until you consider that it is locked in a 1:1 orbit with the Earth."[13] Manfred Lachs, who served as the chairman of the Legal Sub-Committee of the UN COPUOS during the negotiation of the Outer Space Treaty and was the president of the International Court of Justice (ICJ), aptly observed that there are "myriads of celestial bodies-from giants

---

[11] *Official Records of the General Assembly, Twenty-First Session, First Committee,* Summary Records of Meetings, 1492nd Meeting, 17 December 1966, UN Doc. A/C.1/ SR. 1492, pp. 427- 428. (Emphasis added). Similarly, the Soviet delegate stated that Article I, Paragraph 1, was not "a mere statement of the rights of States" but was designed "to guarantee that the interests, not only of individual States, but of all countries and of the international community as a whole, would be protected." UN. Doc. A/A C. *1* 05/C. 2/SR. 57 (20 October 1966), at 12.

[12] Jerry Coffey, "*CELESTIAL BODY," UNIVERSE TODAY,* 27 December 2009, available online at: http://www.universetoday.com/48671/celestial-body/

[13] *Ibid.*

to micro-meteorites, yet size could not be accepted as a test for their legal status" and submitted that for "the present, the term 'celestial bodies' as employed in the relevant instruments should therefore be viewed as the largest common denominator of all 'land areas' in outer space."[14] Any classification of celestial bodies according to size for the purpose of determining whether these bodies fall under the scope of existing international space law would be arbitrary, controversial and legally untenable. Thus, in summary, all planets, comets, stars, asteroids, and meteorites, irrespective of their sizes, forms and orbits, are celestial bodies and must be considered regulated under international space law, including the Outer Space Treaty and the Moon Agreement.

The second paragraph of Article I of the Outer Space Treaty reflects the fundamental principle of freedom of exploration and use of outer space and is also known as the "freedom principle." Such freedom was designed and has been recognized to be broad in nature and scope since the objective of the Outer Space Treaty was to create an international governance regime for all space activities rather than to regulate any specific one.

In general, many international treaties have been reactive. They are often adopted to resolve certain existing issues and often contain prohibitions or limitations upon specified actions of States. On the contrary, the Outer Space Treaty is proactive in nature. Although two States, the United States and the former Soviet Union, were essentially the only ones involved in space operations at the time, the international community negotiated the treaty well before space activities matured. Therefore, the treaty was designed to contain a few prohibitions, but several freedoms of action and numerous prescriptive obligations for States, with which all States Parties to the treaty must comply. It must be understood that the second paragraph of Article I covers all space activities that were known or unknown at the time of the conclusion of the Outer Space Treaty, whether or not they are specifically and expressly mentioned in the treaty. For example, satellite remote sensing activities were known at the time but not specifically mentioned in the treaty. While the operation of private space stations, active space debris removal and on-orbit servicing of satellites were neither known nor mentioned at the time of drafting, they all fall within the ambit of the treaty, as they can all be encapsulated under the term "space activities."

Since the terms freedom for "exploration" and "use" are not defined in the Outer Space Treaty, they must be interpreted in good faith in accordance with the ordinary meaning to be given to these terms in their context and in the light of the object and purpose of the treaty. Thus they should be understood to mean what they convey in the normal sense, including exploitation of space's natural resources with the application of current and future space technologies. Moreover, being declaratory in nature, the second paragraph of Article I recognizes the freedoms of exploration, use and exploitation of outer space and celestial bodies not only by States Parties to the Outer Space Treaty but by *all* States. In addition, not only the States but also their public and private entities (private corporations) are entitled to such freedoms, though the latter

[14] Manfred Lachs, *The Law of Outer Space: An Experience in Contemporary Law-Making*, 1972, p.46.

can exercise the freedoms contained in the treaty only pursuant to (as determined by) authorization (licensing) and continuous supervision of their respective States.[15] In other words, private corporations, at least those that are under the jurisdiction of States that are party to the Outer Space Treaty, are not permitted to carry out any space activity, including the exploitation of natural resources in space, without the permission of their respective governments.

States may exercise the freedoms guaranteed under the second paragraph of Article I individually or in collaboration with other States or their public or private entities through joint ventures or international intergovernmental organizations.[16] Such international organizations are not able to become party to the treaty or declare their acceptance of the rights and obligations provided therein. However, in all situations the responsibility for compliance with all the provisions of Outer Space Treaty is borne by the concerned States, even when space activities are carried out by international organizations. Article VI of the Outer Space Treaty also imposes responsibility to comply with the provisions of the treaty on the concerned international organization (in addition to the States Parties to the treaty participating in such an organization); however, in practice, it could become a challenge to ensure the compliance of an international organization when the organization is not party to the treaty.

It is imperative to keep in mind that, although the freedom of exploration, use and exploitation is broad in nature and scope, it certainly is not unfettered and absolute and must be exercised within the limits imposed (the parameters prescribed or to the extent allowed) by the Outer Space Treaty and other applicable principles and rules of international law. The second paragraph of Article I of the Outer Space Treaty itself contains three main limits upon or requirements for the exercise of freedoms of exploration, use and exploitation. States must exercise their freedom (1) "without discrimination of any kind," (2) "on a basis of equality" and (3) "in accordance with international law." The phrase "without discrimination of any kind," read in conjunction with the preamble and provisions of the first paragraph of Article I of the treaty, implies that the first explorers, users and exploiters of outer space and the celestial bodies cannot use the lateness of other States as a basis for jeopardizing the freedoms of the latter States to explore, use and exploit outer space. The phrase "on a basis of equality" indicates the legal equality of the right of freedom of all States, which implies equality in law (i.e., *de jure* equality) or "sovereign equality" as recognized in Article 2(1) of the Charter of the United Nations.[17] Finally, the phrase "in accordance with international law" requires compliance with current and future principles and rules of conventional and customary international law as well as the Charter of the United Nations. An interesting example of this is the prohibition of abuse of rights.[18]

[15] *Outer Space Treaty*, Art. VI.

[16] *Outer Space Treaty*, Art. XIII.

[17] *Charter of the United Nations*, 26 June 1945, CAN TS 1945 No 7.

[18] According to Alexandre Kiss, under "international law, abuse of rights refers to a State exercising a right either in a way which impedes the enjoyment by other States of their own rights or for an end different from that for which the right was created": Alexandre Kiss, "Abuse of Rights," 2006, available online at Oxford Public International Law: http://opil.ouplaw.

The other provisions of the Outer Space Treaty set out three specific limits on the exercise of the freedoms guaranteed in the second paragraph of Article I as follows: (a) the requirement to comply with the common interest principle (as discussed above); (b) the prohibition of national appropriation under Article II (as discussed below) and (b) the requirement to pay "due regard to the corresponding interests of all other States Parties to the Treaty" as incorporated in Article IX. Any exercise of the freedoms under the second paragraph of Article I by States, or by the entities for which they are responsible, outside or contrary to the above-mentioned (and other applicable) limits or requirements will be unlawful from an international law perspective. Any such illegal action of a State Party to the Outer Space Treaty may give rise to an international conflict and thus could further put the State in violation of its obligation specified in Article III; i.e., duty to carry on its space activities "in the interest of maintaining international peace and security and promoting international co-operation and understanding."[19] Violation of any international obligation entails international responsibility under international law.

Prohibition of Appropriation of Outer Space and Celestial Bodies

While the first and second paragraphs of Article I contain a prescriptive provision and a freedom respectively, Article II of the Outer Space Treaty sets out one of the few prohibitions found in the treaty. It provides that outer space, including the Moon and other celestial bodies, is not subject to national appropriation by claim of sovereignty, by means of use or occupation, or by any other means.

This article contains the so called "non-appropriation principle" and is complementary to and a necessary prerequisite for, the common interest and freedom principles. Outer space and the celestial bodies simply cannot be explored and used freely by, and for the benefit of, all countries if some are allowed to appropriate these commons of humankind. The underlying rationale and purpose of Article II were expressed in a statement made by Herbert Reis, the U. S. representative to the UN COPUOS during the second session of the Working Group on Direct Broadcast Satellites meeting held on 31 July 31, 1969:

> *The negotiating history of the [Outer Space] Treaty shows that the purpose of this provision [Article II] was to prohibit a repetition of the race for the acquisition of national sovereignty over overseas territories that developed in the sixteenth, seventeenth, eighteenth and nineteenth centuries. The Treaty makes clear that no user of space may lay claim to, or seek to establish, national sovereignty over outer space or a celestial body.*[20]

---

com/view/10.1093/law:epil/9780199231690/law-9780199231690-e1371 Admitting that the prohibition of abuse of rights is "problematic because of differences in the content of the concept itself" Kiss asserts that the "idea that a subject [State] of rights and competences can misuse them seems to be inherent to legal thinking and to have roots in all legal systems and leads to the establishment of controls on the use of recognized rights." Ibid.

[19] *Outer Space Treaty,* Art. III.

[20] Cited from Erik N. Valters, 'Perspectives in the Emerging Law of Satellite Communications' (1970) 5 Stanford Journal of International Studies 53, at 66 - 67. Also cited in Kathryn M. Queeney, *Direct Broadcast Satellites and the United Nations*, BRILL, 1978, p. 54.

This statement reflects the strong desire of the negotiating States to discontinue the traditional practice of extending national sovereignty to other territories that on Earth and throughout human history has resulted in the brutal colonization (and deaths) of millions of people, countless tragedies, numerous and ferocious wars, the spread of diseases, degradation of the environment, and the reckless exploitation, and even depletion, of natural resources. This also clearly explains the rationale for incorporating a very broad prohibitive provision against appropriation in Article II.

Article II prohibits "national" appropriation of outer space and celestial bodies but is silent on the question of appropriation by individuals or private entities. According to Stephen Gorove "an individual acting on his own behalf or on behalf of another individual or a private association or an international organization could lawfully appropriate any part of outer space."[21] Such interpretation of Article II is illogical and untenable.[22] As noted above, States are under obligation to ensure compliance with the provisions of the Outer Space Treaty by their private entities. If private appropriation were permitted to appropriate outer space and celestial bodies, it would defeat the purpose of the treaty and nullify the common interest and freedom principles. Moreover, the negotiating history of the Outer Space Treaty clearly indicates that both public and private property rights are not allowed in outer space and on celestial bodies. For example, on August 4, 1966, the Belgian representative to UN COPUOS noted that the term "non-appropriation," advanced by several delegations—apparently without contradiction by others—covered both the claims of sovereignty and "the creation of titles to property in private law."[23] Similarly, on December 17, 1967, the French delegate to UN COPUOS stressed that a basic principle of the 1967 Outer Space Treaty was that there was a "prohibition of any claim to sovereignty or property rights in space."[24] According to Manfred Lachs, the celestial bodies "cannot be subjects of proprietary rights."[25] Similarly, Dr. Manfred A. Dauses is of the view that the "term 'appropriation by means of use' may be interpreted as the establishment of exclusive rights over certain uses of particular segments of space or celestial bodies, such as exclusionary rights of way or the monopolistic exploitation of cosmic resources."[26]

It is clear that the non-appropriation principle prohibits the appropriation of outer space by claim of sovereignty, by means of use or occupation, or *by any means whatsoever*. This intentionally designed provision is very broad and makes it absolutely clear that traditional means of acquisition of territory or any other means will not be a justification for appropriation of outer space, celestial bodies or parts thereof. The meaning of the concept of "use" that appears

[21] Stephen Gorove, *Interpreting Article II of the Outer Space Treaty*, 37 Fordham L. Rev. 349 (1969), at 351.

[22] Ram Jakhu, "Legal Issues Relating To the Global Public Interest in Outer Space," 32 *Journal of Space Law* (2006) 31, at 44-46.

[23] Cited in Carl Christol, *Article* 2 *of the* 1967 *Principles Treaty Revisited,* IX (1984) Annals of Air and Space Law, 217, at 236.

[24] *Ibid,* at 218.

[25] Manfred Lachs, *supra* note 14.

[26] Manfred A. Dauses, *THE RELATIVE AUTONOMY OF SPACE LAW*, available online at: https://opus4.kobv.de/opus4-bamberg/files/6652/The_Relative_Autonomy_of_Space_LawOCRseA2.pdf.

within the non-appropriation principle as prescribed in Article II of the Outer Space Treaty must be considered in light of the terms of Article I, which denotes the principle of the freedom of "use" and must naturally encompass the exploitation of the natural resources of outer space and celestial bodies. It would appear that although the "use" of outer space is permitted for all States, including their governmental and non-governmental entities, under no circumstances would any amount of such "use" ever suffice to justify a claim of sovereignty or ownership over the whole or any part of outer space, including the Moon and other celestial bodies. This construction of the meaning of Articles I and II of the Outer Space Treaty implies that the exploitation of the natural resources of the Moon and other celestial bodies constitutes a use of outer space that is contemplated by the freedom principle specified in the Outer Space Treaty … . However, for the purposes of Article II of the Outer Space Treaty, this use does not, and can never be such as to constitute a "national appropriation" [of outer space] giving rise to ownership rights.[27]

Since the adoption of the Outer Space Treaty, several attempts have been made to appropriate or assert ownership rights in outer space or over celestial bodies. The first attempt to appropriate a part of outer space was made in 1976. Eight equatorial countries (namely, Brazil, Colombia, Congo, Ecuador, Indonesia, Kenya, Uganda, and Zaire), under the so-called Bogota Declaration, asserted that "the segments of geostationary synchronous orbit are part of the territory over which Equatorial states exercise their national sovereignty."[28] The international community rejected this proclamation as being contrary to the Outer Space Treaty, particularly its Article II.[29] It should be noted that the three branches of the U. S. government (i.e., the Legislative,[30] Executive[31] and

[27] Steven Freeland and Ram S Jakhu, "Commentary on Article II of the Outer Space Treaty" in Stephan Hobe, Bernhardt Schmidt-Tedd & Kai-Uwe Schrogl, eds, *Cologne Commentary on Space Law* Vol. 1 (Cologne: Carl Heymanns Verlag, 2010) at 53.

[28] *Declaration of the First Meeting of Equatorial Countries*, adopted and signed in Bogota by the Heads of Delegations on December 3, 1976. The text of the Declaration is available online at: https://bogotadeclaration.wordpress.com/declaration-of-1976/

[29] Ram Jakhu, "The Legal Status of the Geostationary Orbit", VII *Annals of Air and Space Law*, 1982, pp. 333-352.

[30] Under section 403 of the *Space Resource Exploration and Utilization Act of 2015* (51 USC Title IV, Pub. L. 114-90; Bill HR 2262), which was signed by President Barack Obama on 25th November 2015, the U.S. Congress issued a disclaimer of extraterritorial sovereignty expressing that "by the enactment of this Act, the United States does not thereby assert sovereignty or sovereign or exclusive rights or jurisdiction over, or the ownership of, any celestial body."

[31] On 16 February 2000, Mr. Gregory Nemitz (Chief Executive Officer of Orbital Development of San Diego, USA) sent a claim to U.S. National Aeronautics and Space Administration (NASA) for payment of a $20 "parking/storage fee" for NASA's Near Earth Asteroid Rendezvous Shoemaker spacecraft that landed on the asteroid 433 Eros on the basis that "since March 3, 2000 Orbital Development has owned Eros by virtue of a property claim filed on that date with an organization called Archimedes Institute." (See: http://www.orbdev.com/010309.html) Refuting such a claim, Mr. Edward A. Frankle, the General Counsel with NASA, responded on 9th March 2001 to Mr. Nemitz by expressing NASA's position that "the Article II of the Outer Space Treaty of 1967, to which the United States is a party, states: "Outer space,

Judiciary[32]) have declared their acceptance of and respect for the non-appropriation principle, which applies to both public and private entities. Similarly, Chinese[33] and Canadian courts[34] have confirmed the non-appropriative character of outer space, the Moon and other celestial bodies. Freeland and Jakhu aptly conclude that the prohibition of appropriation specified in Article II of the Outer Space Treaty has become a rule of customary international law (in the form of a *jus cogens* norm[35]) and is thus applicable to all States, whether or not they are parties to the Outer Space Treaty. All States are under an obligation not only themselves to comply with the principles set out in Article II, but also to ensure that their respective non-governmental entities do not in any manner or form act contrary to this important legal norm.[36]

Prohibition of Appropriation of Space Natural Resources

The question that arises is whether the prohibition of appropriation of outer space and celestial bodies extends to their natural resources. In other words, is it legitimate for a State, or its public entities and private companies, to exploit, take possession or ownership of natural resources in space to the

including the moon and other celestial bodies, is not subject to national appropriation by claim of sovereignty, by means of use or occupation, or by any other means." 610 U.N.T.S. 205, 18 U.S.T. 2410. If Orbital Development or its principals are U.S. nationals, this treaty provision would seem to preclude any claim to own Eros. Therefore, NASA respectfully declines to make the requested payment at this time." Ibid.

[32] Not being satisfied by NASA's response (*Ibid.*), Mr. Nemitz took his case to the Federal District Court in the State of Nevada, which dismissed his claim of a private property rights on an asteroid by ruling that "neither the failure [by] … the United States to ratify the … Moon Treaty, nor … the Outer Space Treaty, created any rights in Nemitz to appropriate private property rights on asteroids." *Nemitz v. U.S.*, Slip Copy, 2004 WL 316704, D. Nev., 26 April 2004. On appeal, the Ninth Circuit Court of Appeals upheld the ruling of the lower court "for the reasons stated by the district court." Nemitz v. NASA, 126 Fed. Appx. 343 (9th Cir. (Nev.) 10 February 2005).

[33] In 2007, the Beijing First Intermediate People's Court ruled against the Lunar Embassy in China Company, which was selling plots on the Moon, stating that no individual or State could claim ownership of the Moon. In its pronunciation the Court cited the fact that China was a party to the Outer Space Treaty, which prohibits appropriation of outer space and its parts. See: *Court Rejects Lunar Embassy's Right of Moon Land Selling,* Xinhua News Agency March 17, 2007, available online at: http://www.china.org.cn/english/China/203329.htm. Also see:

Beijing authorities suspend license of "Lunar Embassy", November 07, 2005, available online at: http://en.people.cn/200511/07/eng20051107_219609.html

[34] In 2012, Judge Alain Michaud declared Sylvio Langevin, a Quebec man, to be a quarrelsome litigant barring him from filing lawsuits claiming ownership over nine planets, four of Jupiter's moons as well as the space between the heavenly bodies. See: Brian Daly, *Man sues for ownership of most of solar system,* QMI Agency, 1 March 2012, available online at: http://cnews.canoe.com/CNEWS/WeirdNews/2012/03/01/19445846.html

[35] Principle of *jus cogens,* according to Article 53 of the VCLT is "a peremptory norm of general international law is a norm accepted and recognized by the international community of States as a whole as a norm from which no derogation is permitted and which can be modified only by a subsequent norm of general international law having the same character." A treaty is void if, at the time of its conclusion, it conflicts with a norm of *jus cogens.* Ibid.

[36] Steven Freeland and Ram S Jakhu, *supra note 27,* at 63.

exclusion of all other States and their entities? Is the exercise of private property rights over these natural resources compatible with the provisions of the Outer Space Treaty?

Some space law writers[37] and States, particularly the United States, are of the view that Article II does not prohibit exploitation of natural resources. In other words, while the surfaces of the Moon and celestial bodies are subject to the non-appropriation provision of the Outer Space Treaty, their natural resources can be appropriated. This view seems to be the rationale behind the U. S. Space Act of 2015, pursuant to which U. S. citizens (including private corporations) may be authorized to engage in commercial exploration for and commercial recovery of space resources free from harmful interference, [but] in accordance with the international obligations of the United States and subject to authorization and continuing supervision by the federal government.[38]

The national law aspects of the Act are discussed in the next chapter of this book. After analyzing the provisions of this act and its compatibility with the Outer Space Treaty, the Board of Directors of the International Institute of Space Law (IISL) issued a position paper on space resource mining on December 20, 2015.[39] The central point of the IISL position is that "in view of the absence of a clear prohibition of the taking of resources in the Outer Space Treaty one can conclude that the use of space resources is permitted."[40]

It is respectfully submitted that the rationale for and basis of the IISL position do not seem to be fully accurate, and thus its characterizations, conclusions and implications may not be supported by the current *lex lata*. Firstly, the IISL position appears to be based on the concept that under international law "what is not prohibited is permitted." The Permanent Court of International Justice employed this concept *obiter dictum*[41] in the 1927 *Lotus* case.[42] The *obiter dictum* and the reasoning of the PCIJ on the real issue of the *Lotus* have been extensively criticized by scholars and rejected by the ICJ as well as the international community.[43] Moreover, several renowned space law scholars, including Manfred Lachs and Carl Christol, have categorically declared that the rationale of the *Lotus* case (including the above-mentioned concept) is inapplicable to outer space matters.[44]

[37] For example, Tanja Masson-Zwaan, has been reported to be of the opinion that "existing treaties do not seem to prohibit ownership of extracted resources"; in Marcia S. Smith, "Posey, Kilmer Introduce ASTEROIDS Act To Grant Property Rights to Asteroid Resources", 10 July 2014, available online at http://www.spacepolicyonline.com/news/posey-kilmer-introduce-asteroids-act-to-grant-property-rights-to-asteroid-resources

[38] Section 403 of the *Space Resource Exploration and Utilization Act of 2015* (51 USC Title IV, Pub. L. 114-90; Bill HR 2262). For the text of the Act, see Appendix to this book.

[39] The Text of the IISL Position Paper is available at IISL: http://www.iislweb.org/docs/SpaceResourceMining.pdf

[40] *Ibid,* section II.2.

[41] "*Obiter dictum*" is only an observation which is not necessary or important in determining the central issue in a case before a court of law.

[42] *The Case of the S.S. Lotus* (Fr. v. Turk.), 1927 P.C.I.J. (ser. A) No. 10 (Sept. 7)

[43] For details, see Ram Jakhu, *supra* note 22, at 41-42.

[44] *Ibid.*

Secondly, if "the use of space resources is permitted," then the question of the source (legal basis) of that "permission" arises. It is submitted that the freedoms guaranteed in the second paragraph of Article I of the Outer Space Treaty is the source (legal basis) for permitting resource exploitation in outer space. In other words, the freedom to exploit these natural resources is within the Outer Space Treaty and not outside this international legal regime-creating treaty.[45] Thirdly, the Outer Space Treaty uses the terms "exploration and use" but not "exploitation" of outer space and celestial bodies and their natural resources. As indicated above, the "use" of outer space as prescribed in Article I of the Outer Space Treaty encompasses "resource exploitation." If the term "use" is considered to include the exploitation of space's natural resources under Article I, it should also be encompassed under Article II and in turn prohibit the appropriation of these natural resources. Thus, in essence, what the Outer Space Treaty allows "use" of outer space and celestial bodies but prohibits their "appropriate by use".

In practice, it may be difficult and complex to draw a line between legitimate "use" and the prohibited "appropriation by use" of outer space and its natural resources, but such complexity must not nullify the effect of both the legal principles of freedom and non-appropriation. It should be noted that the IISL position is not the firm and final statement of *lex lata* as it aptly acknowledges that in the absence of a clear prohibition of the taking of resources in the Outer Space Treaty, "[T]he new United States Act is *a possible* interpretation of the Outer Space Treaty. *Whether and to what extent this interpretation is shared by other States remains to be seen.*"[46]

The three key legal principles incorporated in the Outer Space Treaty (i.e., common interest, freedom and non-appropriation), which were painstakingly and meticulously negotiated by the international community nearly half a century ago, form a three-dimensional and foundational pillars upon which the entire international legal regime for outer space and celestial bodies has been constructed. Disregard for one of them will certainly result in the weakening of that foundation and consequently in the collapse of the entire global space order in place today. They appear inconsistent, but with a careful examination of their

[45] Some space law writers are of the opinion that space natural resources can be exploited, without the constraints of Article II 'appropriation' based on the analogy with the law of the sea. For example. Fabio Tronchetti is of the view that, "While some authors express the view that the restriction in Article II [of Outer Sapce Treaty] applies equally to outer space and its resources, others, the majority, argue that by analogy with the rules regulating the *freedom of the high seas,* the appropriation of space resources merely forms part of the freedom of exploration and use of outer space. This paper shares the opinion of the second group of authors": See Fabio Tronchetti, "The Moon Agreement In The 21st Century: Addressing Its Potential Role in the Era of Commercial Exploitation of the Natural Resources Of The Moon And Other Celestial Bodies", 36 *Journal of Space Law* (2010) 489, at 498. (Footnotes omitted and emphasis added). This sort of analogies are appropriately refuted by Philip de Man, See Philip de Man, *Exclusive Use in an Inclusive Environment: The Meaning of the Non-Appropriation Principle for Space Resource Exploitation,* (2016), pp. 15-26 (Forthcoming publication in Springer: The Space Regulations Library Series).

[46] The IISL Position Paper, *supra* note 39, section II.2.

provisions, carried out strictly in accordance with the internationally agreed upon rules of treaty interpretation, a fair and logical balance of different provisions, interests and perspectives can and should be understood and achieved.

It is therefore submitted that based on a combined reading of the provisions of Articles I and II of the Outer Space Treaty, commercial exploitation of natural resources in space would constitute appropriation unless carried out in compliance with the requirements not only of Article I but also of other provisions of the Outer Space Treaty. In other words, the commercial exploitation of these natural resources is permitted when carried out in compliance with the requirements of Article I and other provisions of the Outer Space Treaty. Freeland and Jakhu note a situation where the exploitation of the natural resources of a celestial body—say, a small asteroid—is of such a scale that, in effect, the celestial body is exploited "out of existence" … could be unlawful, since it would, in all likelihood, violate other principles of international space law, such as the requirement that it be "for the benefit and in the interests" of all States' and that due regard is to be paid to the corresponding interests of all other States.[47]

It can, therefore, be concluded that compliance of the U. S. Space Act of 2015 with the provisions of the Outer Space Treaty will depend upon the legislative and regulatory steps that the U. S. government would take in the implementation of this act through its national regulatory system. The act requires the U. S. government to comply with its international obligations, which would logically include its obligations under Articles I, II and IX of the Outer Space Treaty since the United States is party to that treaty. If by legislative fiat, private U. S. corporations are entitled to exclusive property rights over space natural resources without due regard to these and other applicable provisions of the Outer Space Treaty, the U. S. action might be considered to be in violation of its international obligations.

It is also relevant here to discuss the legal status of natural resources in space, which in turn is based on whether or not outer space constitutes a global commons. The term "global commons" is not a legal concept but a political one that is primarily used by writers, journalists and politicians for advocating their respective perspectives and policies. In common parlance, the "global commons" generally refers to something that is common to all humanity and not subject to national sovereignty. A typical example of a global commons is the high seas, which, we must hasten to add, is governed by a specific international legal regime.[48] This broad and undefined term is without any legal origin, basis and context, and it does not precisely determine the *lex lata* of the status of natural resources in space, which should be decided exclusively on the basis of applicable international space law, particularly the Outer Space Treaty and the Moon Agreement. Manfred Lachs' views on this issue are quite pertinent, for he cautioned "the 'mechanical transfer' of institutions

[47] Steven Freeland and Ram S. Jakhu, *supra* note 27, at 53.

[48] *United Nations Convention on the Law of the Sea, 10 December 1982, 1833 UNTS 3 (entered into force 16 November 1994).*

from one environment to another is of little avail: it may lead to distortions and even seriously stunt the development of the new branch of law [like space law]."[49] However, such terms as global commons may serve as models for the development of new space law treaties (*lex de ferenda*).

The 1979 Moon Agreement

The Moon Agreement was intended to initiate discussions on, leading to the eventual elaboration of, a detailed international regime mainly for the exploration and exploitation of the natural resources of the Moon and celestial bodies when it is about to become feasible. In this regard, the Moon Agreement builds upon the provisions of the Outer Space Treaty insofar as they relate to the use of the resources of the Moon and other celestial bodies.

The terms of the Moon Agreement are explicitly more conducive to the exploitation of the natural resources of the Moon and other celestial bodies. First, Article 6(2) of the Moon Agreement explicitly grants to States' parties the right to collect and remove from the Moon samples of mineral and other substances during the conducting of scientific investigations. Further, in the course of their scientific investigations, States' parties may also use mineral and other substances of the Moon in quantities appropriate in support of their missions. Secondly, although Article 11(2) of the Moon Agreement replicates the prohibitions found in Article II of the Outer Space Treaty, it is believed that one of the principal objects of the Moon Agreement is to promote the "exploitation" of the natural resources of the Moon through its own provisions and provide for the eventual establishment of a dedicated international regime to facilitate such exploitation.[50] The Moon Agreement is an improvement upon, and being later in time may likely prevail over, the provisions of the Outer Space Treaty for States that are parties to both these treaties.

It is clear therefore that the prohibition of national appropriation of outer space as prescribed in Article II of the Outer Space Treaty and Article 11(2) of the Moon Agreement does not in and of itself restrict the exploitation of the natural resources of the Moon and other celestial bodies, an activity that will involve the removal of such resources from the Moon and other celestial bodies. Rather, the principle is directed towards preventing claims to ownership rights over outer space, celestial bodies or portions thereof. Indeed, this position is reinforced by the terms of Article 11(3) of the Moon Agreement according to which the placement of *inter alia* personnel, facilities or installations on or below the surface of a celestial body shall not create a right of ownership therein. In fact, similar to the provisions of Article II of the Outer Space Treaty and Article 11(2) of the Moon Agreement, the terms used in Article 11(3) of the Moon Agreement suggest that the prohibition against the appropriation of outer space only extends to natural resources while they are "in place" in outer space. Once the natural

[49] Manfred Lachs, *supra* note 14, p.21.

[50] Steven Freeland and Ram S Jakhu, *supra* note 27, at 59.

resources have been removed or extracted from their original locale on the Moon or a celestial body, there is no provision that prohibits the exercise of ownership or property rights over them, if however, such removal or extraction is in compliance with the applicable principles and rules of international space law. In this respect, the provisions of the Moon Agreement are more specific and beneficial to any country the private sector entities of which have an interest in commercial exploitation of the resources of outer space.[51] Freeland and Jakhu therefore argue the prohibition against appropriation would not prevent public or private entities from receiving—under an international regime to be established in the future—what might be termed "extra-terrestrial exploitative rights" in relation to the natural resources of outer space.[52]

Article 11(1) of the Moon Agreement provides that "[t]he Moon and its natural resources are the common heritage of mankind *which finds its expression in the provisions of this Agreement, in particular in paragraph 5 of [Article 11]*".[53] Under Article 11(5):

> *States Parties to [the Moon] Agreement ... undertake to establish an international regime, including appropriate procedures, to govern the exploitation of the natural resources of the Moon as such exploitation is about to become feasible. This provision shall be implemented in accordance with article 18 of [the Moon] Agreement.*[54]

It appears that many countries cite the inclusion of the common heritage of mankind concept within the Moon Agreement as the reason why they are not interested in becoming parties thereto. In support of their position, these countries refer to the unsuccessful application of the common heritage of mankind concept in international legal regimes governing the terrestrial environment and also in resource exploitation regimes established for Antarctica and the deep seabed.

However, it is significant to note that the language of Article 11(1) of the Moon Agreement does not indicate a wholesale importation of the common heritage of mankind concept into international space law. Rather, the Moon Agreement makes an important distinction as to what the concept specifically entails and how it shall be applied in the context of the natural resources of the Moon and other celestial bodies. Most importantly, the meaning and effect of the common heritage of mankind concept as used in the Moon Agreement is that it simply requires States to develop and establish a mutual international legal regime, including appropriate procedures, to govern the exploitation of the natural resources of the Moon, as such exploitation is about to become feasible. The concept may be included

[51] It is important to note in this regard that the Outer Space Treaty uses the terms "exploration and use" but not "exploitation" of outer space. As indicated above, it is generally agreed that the "use" of outer space as prescribed in Article I of the Outer Space Treaty presumably encompasses "resource exploitation". However, it is only the Moon Agreement that specifically uses the term "exploitation". States parties to the Moon Agreement therefore have a more explicit basis for asserting their right to exploit the natural resources of outer space.

[52] Steven Freeland and Ram S. Jakhu, *supra* note 27, at 60.

[53] *Moon Agreement*, Art. 11(1) [emphasis added].

[54] *Moon Agreement*, Art. 11(5).

in the proposed international regime, which may or may not be established, if the negotiating States so wish. It is unfortunate that some States and many space law writers often mistakenly overplay the importance of the concept of common heritage of mankind. This concept has very little, if any, role in the regulatory regime established under the Moon Agreement.

Although the Moon Agreement sets out certain overarching objectives that must be achieved during the development and establishment of the envisaged international regime to govern the exploitation of the natural resources of the Moon, and arguably of other celestial bodies,[55] it does not restrict States to a specified mechanism for the fulfillment of this requirement. In this regard, the Moon Agreement differs significantly from the Antarctic Treaty or the 1982 U. N. Convention on the Law of the Sea and its 1994 agreement relating to the implementation of Part XI thereof. As such, States' parties to the Moon Agreement are at complete liberty to determine which regulatory model provides the best medium for managing the exploitation of the exhaustible natural resources of the Moon. It is therefore submitted that those States that desire to have a say in the establishment of a regime to govern the exploitation of the natural resources of the Moon and other celestial bodies should accede to and ratify the Moon Agreement without further delay.

[55] For instance, Article 11(7) of the Moon Agreement sets out the main purposes of the international regime to be established as including the following:

(a) The orderly and safe development of the natural resources of the Moon;
(b) The rational management of those resources;
(c) The expansion of opportunities in the use of those resources;

(d) An equitable sharing by all States Parties in the benefits derived from those resources, whereby the interests and needs of the developing countries, as well as the efforts of those countries which have contributed either directly or indirectly to the exploration of the Moon, shall be given special consideration.

Conclusion

As things stand now, if China, India, or Russia, for example, were successful in rallying several developing countries to ratify the Moon Agreement, the resulting reality would be that many of the current spacefaring nations that have not ratified the Moon Agreement (including the United States) would seemingly be excluded from the deliberations leading to the establishment of an international regime to govern the exploitation of space natural resources. Furthermore, the parties to the agreement that end up on the Moon or any other celestial body would have a more explicit basis (i.e., the Moon Agreement) to assert their right to exploit the natural resources present therein as compared to those countries that are not yet party to the Moon Agreement.

It should be noted that the Moon Agreement offers some advantages that are not available under the Outer Space Treaty.[56] For instance, Article 3(4) of

[56] For detailed discussions, see Joint Statement on the benefits of adherence to the Agreement

the Moon Agreement expressly prohibits the establishment of military bases on the Moon and other celestial bodies. More importantly, Article 3(2) declares that any threat or use of force or any other hostile act or threat of hostile act on the Moon is illegal. Such threat or act cannot be committed in relation to Earth, the Moon, a spacecraft, the personnel of spacecraft or manmade space objects, including those on the Moon and other celestial bodies. Such an unequivocal prohibition of threat or use of force on the Moon and other celestial bodies is not found in the Outer Space Treaty. Thus, the Moon Agreement establishes the rule of law in connection with the exploration of the Moon and other celestial bodies under an exclusively peaceful and threat-free environment. These conditions are believed to be an important factor (inducement) for attracting the necessary financial investments required for resource exploration and exploitation ventures in space.

Thus, for the moment, the 1979 Moon Agreement constitutes an important international legal document. The lack of signatures ratifications of to this document, and the possibility that national laws might be enacted that provide contradictory interpretations of that agreement, cast some doubt as to what the status of international law is with regard to space mining. It will probably be some years before the situation is clarified. Some of the key issues to bear in mind and monitor are the following:

(a) To what extent does the current situation lead countries to ratify the Moon Agreement or, in fact, result in countries explicitly indicating that they will not sign on to it?
(b) To what extent do nations, and particularly spacefaring nations, move to enact national legislations of the type and scope of which is parallel to that adopted and signed into law in the United States?
(c) To what degree do UN COPUOS, the International Space Exploration Coordination Group (ISECG), the International Astronautics Federation, and others, provide useful and successful means to discuss the future of space mining and resource utilization in outer space and to help coordinate the activities of countries and private enterprise seeking to expand their activities in outer space?

Governing the Activities of States on the Moon and Other Celestial Bodies of 1979 by States Parties to that Agreement; Committee on the Peaceful Uses of Outer Space Legal Subcommittee, Forty-seventh session; UN Doc A/AC.105/C.2/2008/CRP.11 of 2 April 2008; see also Ram Jakhu and Maria Buzdugan, "The Role of Private Actors: Commercial Development of the Outer Space Resources, Including Those of the Moon and other Celestial Bodies: Economic and Legal Implications," 6 *Astropolitics*, (2008), pp. 201, at 221 *et seq.;* Vid Beldavs, "The International Lunar Decade", *The Space Review,* 13 January 2014,online: The Space Review http://www.thespacereview.com/article/2431/1.

11 National Space Laws and the Exploitation of Natural Resources from Space

Article VI of the 1967 Outer Space Treaty obliges States' parties to bear international responsibility for national activities in outer space, whether carried out by governmental agencies or non-governmental entities (including private corporations).[1] It further specifies that the activities of non-governmental entities in outer space, including the Moon and other celestial bodies, shall require authorization and continuing supervision by the appropriate State party to the treaty. In line with these international obligations, many countries have enacted national laws that require non-governmental entities to obtain some form of governmental authorization before they engage in space activities. As a result, there has emerged in many countries—a specific body of law that governs the conduct of space activities by governmental and non-governmental entities that may be characterized as national space law. Professor Paul Dempsey of McGill University has provided one of the most recent and comprehensive reviews of national space legislation around the world.[2]

Each country enacts national space law(s) to regulate the conduct of space activities for its own specific reasons. However, the fact remains that the most important reason and the single common basis for doing so is a State's international responsibility for its national activities in outer space as specified in the Outer Space Treaty and in other instruments. Typically, the national space laws of many countries tend to be reactive rather than proactive. The law tends to follow and lag significantly behind developments in the exploration and use of outer space.

Many countries have not enacted any national space laws simply because there is no perceived need for such laws due to the non-existence of space activities. There are several spacefaring

[1] *Treaty on Principles Governing the Activities of States in the Exploration and Use of Outer Space, including the Moon and Other Celestial Bodies*, 27 January 1967, 610 UNTS 205 [*Outer Space Treaty]*, Article VI.

[2] See Paul Stephen Dempsey, "National Legislation Governing Commercial Space Activities," *Journal of Space Safety Engineering*, Vol 1, No 2, December 2014, 44-60.

© Springer International Publishing Switzerland 2017

R.S. Jakhu et al., *Space Mining and Its Regulation*,
Astronautical Engineering, DOI 10.1007/978-3-319-39246-2_11

nations that also do not have basic overarching national space legislation that governs the overall conduct of space activities. Although, in recent times, the trend appears to be changing for the better, the usual situation in many spacefaring countries is that space activities are nationally regulated on a piecemeal basis, with legislation addressing specific sectors of space exploration and use (such as telecommunications, remote sensing, global positioning and satellite navigation) as and when the need arises. The upshot of the foregoing is that, since the private sector in many spacefaring countries has not, until recently, pursued the exploitation of natural resources in space as a commercial space activity, there has been no perceived need among policymakers and lawmakers at the national level to enact statutes and regulations to govern the conduct of such activities. Many countries therefore do not have any regulatory frameworks in place to specifically address the exploitation of the natural resources of outer space by the private sector.

Influenced by "laissez-faire" doctrine, the private sector prefers to leave the exploitation of "new areas" such as outer space, the Arctic or the deep seabed to the first comers and is reluctant to see the development of any rules that may impose restrictions on resource exploitation in such new areas.[3] However, the exploitation of natural resources in space in the absence of clear international and national governance and regulatory regimes might produce counter effects including but not limited to the possibility of conflicting interests among States and private sector entities; a multiplication of claims over the resources; the non-existence of any legal assurance or guarantee for investors that they may be able to recoup their investments in space or at least seek redress via a strong legal framework; the existence of potential international conflicts; and, risks related to pollution and safety and affecting the sustainable development of outer space. Finally, it might prevent scientists from freely doing their work, thereby disabling them from responding to issues relevant to the global public interest and the right of future generations to access natural resources in space.

With the technological breakthroughs recorded in recent years, private sector actors—mainly led by wealthy

[3] See for example the website of the Space Settlement Institute, a non-profit organization established to help promote the human colonization and settlement of outer space http://www.space-settlement-institute.org/. The Space Settlement Institute believes that *private industry*, not government, must assume the lead in space settlement efforts. Accordingly, its mission includes:

- Identification of financial and other incentives to motivate private industry to fulfill such a role;
- Removal of regulatory, legal, and psychological barriers to private sector efforts in space.

The Institute is of the view that a "Lunar Land Claims Recognition Law" that would recognize the right of private lunar settlements to claim and resell the land around their lunar bases is the necessary first step to incentivize permanent human habitation on the Moon. The Institute therefore intends to persuade the US Congress to enact a Space Settlement Prize Act, a draft of which appears on the Institute's website. See also: Alan Wasser & Douglas Jobes "Space Settlements, Property Rights and International Law: Could a Lunar Settlement Claim the Lunar Real Estate It Needs to Survive?" (2008) 73:1 *Journal of Air Law and Commerce* 37.

Americans—are in the process of developing and initiating commercial resource exploitation projects in space, as discussed in Chap. 6. The motivation behind this is the "high frontier," some of them believing that they can even create a real business. Several actors have openly rejected the current legal framework of international space law.[4] Peter Diamandis, the creator of the X-Prize, for instance, is reported to have stated that: "ownership [of property rights on the Moon and other celestial bodies] will be the only powerful driver to open our frontier."[5] The strongest opponent of the present legal framework is Dennis Hope, who has created a website to commercialize parcels of the Moon.[6] For him, the non-appropriation principle of the Outer Space Treaty is only applicable to governments and not to private entities. He did not hesitate to make a declaration to the United Nations in 1980, claiming ownership of the Moon. Despite the fact that the United Nations simply ignored his claims as being untenable and based on a gross misapprehension of the law, Dennis Hope considers his ongoing Internet-based business of selling plots on the Moon to individuals as a legal and legitimate business.

As indicated above, national space law is typically reactive rather than proactive. Despite the recent occurrence of several technological breakthroughs and business developments emanating particularly from the private sector that strongly suggest that the exploitation of the natural resources of outer space is about to become feasible, there has not been much effort at the national level to enact specific regulatory regimes to govern the conduct of such exploitive activities. The fact remains, however, that a number of countries have enacted comprehensive national space laws that are applicable to all kind and manner of space activities. In many other countries, the existing national space laws are fragmented and address different types of space activities and space applications in a piecemeal fashion. Worse still, in many other countries, particularly non-spacefaring nations, there are no space laws in existence.

In the absence of comprehensive/all-encompassing national space laws or specific regulatory regimes dedicated to the governance of resource exploitation activities in outer space, it is not farfetched to presume that most countries, whether States' parties to the Moon Agreement[7] or not, will attempt to regulate resource exploitation activities carried out in outer space on the basis of pre-existing licensing regimes for other types of space activities. This presumption is based on the terms of

[4] See for example: Rand Simberg, *Homesteading the Final Frontier – A Practical Proposal for Securing Property Rights in Space* (Competitive Enterprises Institute, Issue Analysis 2012 No. 2, April 2012) online: http://cei.org/sites/default/files/Rand%20Simberg%20-%20Homesteading%20the%20Final%20Frontier.pdf.

[5] See "Law Journal Article Exposes A Growing Scam: People Getting Rich Selling Deeds To Lunar Real Estate" 2 June 2008, online PR Web, http://www.prweb.com/releases/2008/06/prweb982824.htm.

[6] "What's It All About?" online: Lunar Embassy, http://lunarembassy.com/about.

[7] *Agreement Governing the Activities of States on the Moon and Other Celestial Bodies*, 18 December 1979, 1363 UNTS 3. For the text of the Agreement, see Appendix to this book.

above-mentioned Article VI of the Outer Space Treaty. Further, Article VII of the Outer Space Treaty imposes international liability upon a state that launches or procures the launching of a space object or uses its territory or facilities for the launching of a space object that causes damage to another State or its subjects.

In this chapter we take a look at the national space laws of a handful of spacefaring nations in an effort to determine the existence or non-existence of dedicated regulatory regimes that govern the exploitation of the natural resources of outer space.

The United States

The United States is a State party to the 1967 Outer Space Treaty but not to the 1979 Moon Agreement. It is interesting to note that in 1979, the U. S. Department of State and the American Bar Association suggested that the U. S. Senate should consent to and approve the ratification of the Moon Agreement by the United States. However, many have stated categorically that the U. S. government does not and will never support the Moon Agreement.

In any event, government regulation of space activities in the United States follows the piecemeal approach, with different aspects of private sector space activities being regulated under different pieces of legislation and by different U. S. government agencies, whereas governmental space activities (such as those carried out by NASA and the U. S. Department of Defense) are not subject to regulation by other government agencies. For instance, pursuant to the Communications Act of 1934 as subsequently amended,[8]the Federal Communications Commission (FCC) is responsible for regulating the use of the radio frequency spectrum in the United States, and for granting licenses for telecommunication satellites and Earth stations serving the United States. Its mandate includes assigning satellite orbital locations in accordance with the rules of the International Telecommunication Union. There are other pieces of legislation that address other aspects of space exploration and use in the United States. They have been re-codified under Title 51 of the United States Code.[9]

The Federal Aviation Administration (FAA) regulates commercial space launch activities in the United States (i.e., launching and re-entry of space objects from a U. S. territory or with U. S. facilities).[10] Among other things, Title 51 authorizes the FAA to license launch vehicles; the re-entry of space objects as well as the operation of launch or re-entry sites in the United States. Under Title 51, the FAA has authority to regulate the commercial space transportation industry, to the extent necessary, to ensure compliance with international obligations of the United States and to protect the public health and safety, safety of property, and national security and foreign policy interests of the United States.

Upon the issuance or transfer of a launch or re-entry license, the licensee is required by law to obtain liability

[8] *Communications Act of 1934*, as amended and updated by the *Telecommunications Act of 1996*, Pub L 104-104, 110 Stat 56 (1996).

[9] *National and Commercial Space Programs*, 51 USC (Pub. L. 111–314, § 3, 18 December 2010, 124 Stat 3328).

[10] 51 USC Ch 509.

insurance or demonstrate financial responsibility in amounts to compensate for the maximum probable loss from claims by: (a) third parties for death, bodily injury, or property damage or loss resulting from an activity carried out under the license (limited to U. S $500,000,000 per launch or re-entry, adjusted for inflation); and, (b) the U. S. government against a person for damage or loss to government property resulting from an activity carried out under the license (limited to U. S. $100,000,000, adjusted for inflation). The specified limits are automatically decreased to the maximum liability insurance available on the world market at reasonable cost if, at any point in time, it is impossible to obtain liability insurance coverage up to the specified levels on the world market.

As part of the licensing process, the FAA is entitled to carry out several extensive investigations and reviews, including policy review, safety review, payload review and environmental review. In February 2015, it was reported that the FAA would authorize Bigelow Aerospace to set up an inflatable station on the Moon in exercise of the former's "payload review" authority. Some have observed that the FAA does not specifically possess such power to authorize the conduct of space activities that transcend space transportation.[11] Moreover the U. S. State Department has expressed the view that "the national regulatory framework, in its present form, is ill-equipped to enable the U. S. government to fulfill its obligations" under the 1967 Outer Space Treaty.[12]

Interestingly, on July 20, 2011, NASA published a document entitled: "Recommendations to Space-Faring-Nations: How to Preserve the Historic and Scientific Value of U. S. Government Lunar Artifacts."[13] According to the document, "NASA recognizes the steadily increasing technological capabilities of spacefaring commercial entities and nations throughout the world, and further recognizes that many are on the verge of landing spacecraft on the surface of the Moon. In the 50 years since the first lunar missions, the spaceflight community has not formally provided recommendations to the next generation of lunar explorers on how to preserve the original artifacts and protect ongoing science from the potentially damaging effects of nearby landers."[14] Pending the development of more formal guidance—perhaps through a multilateral approach that reflects the views of various nations on the scientific and historic value of lunar artifacts—NASA has taken the lead to prepare these recommendations on the basis of the collective technical knowledge of its personnel.

According to NASA, the recommendations do not represent mandatory U. S. or international requirements; rather, they are offered to inform lunar spacecraft mission planners interested in helping preserve and protect lunar historic artifacts and potential science opportunities for future missions. The

[11] Irene Klotz, "The FAA: regulating business on the moon," 3 February, 2015, online: http://mobile.reuters.com/article/idUSKBN0L715F20150203?irpc=932.

[12] *Ibid.*

[13] For a copy of the Recommendations, see NASA website: http://www.nasa.gov/directorates/heo/library/reports/lunar-artifacts.html.

[14] *Ibid*, 5.

recommendations are intended to apply to U. S. government artifacts on the lunar surface. By way of example, the NASA recommendations prescribe a descent/landing boundary for U. S. government heritage lunar sites, defined as the outer perimeter that establishes an exclusion radius for the approach path of any lander/spacecraft toward any U. S. government heritage lunar artifacts. For heritage lander sites (e.g., Apollo, Surveyor), this outer perimeter covers an area beginning at the lunar surface site of interest and extending to a 2-km radial distance from the site within which no overflight of a landed spacecraft may occur. For heritage impact sites (e.g., Ranger, S-IVB), the boundary covers an area beginning at the lunar surface site of interest and extending to a 0.5-km radial distance from the center of the impact site, and no overflight of a landed spacecraft may occur within that radius.[15]

NASA insists that these recommendations are consistent with international law, including the 1967 Outer Space Treaty.[16] By putting them forward, NASA claims that it is seeking to promote the development and implementation of appropriate recommendations with interested private sector entities and, as appropriate, working within the U. S. government and with foreign governments. Clearly, the NASA recommendations are unilateral and are explicitly non-binding. More so, their acceptance within the international community is untested, since no space-faring nation has announced plans to approach U. S. lunar artifacts subsequent to their publication. At the very least, the publication of those recommendations demonstrates the fact that the U. S. government acknowledges the renewed interest (particularly from the private sector and also from States that have traditionally not been active in space) in the exploration and eventual exploitation of the natural resources of outer space, and its concern about the impact that this may have on the so called U. S. heritage lunar artifacts.

In July 2014, a bill titled the "American Space Technology for Exploring Resource Opportunities in Deep Space (ASTEROIDS) Act"[17] was

[15] *Ibid.*

[16] In this regard, NASA specifically identifies the following principles contained in the Outer Space Treaty as being relevant:

- That outer space shall be free for exploration and use by all states;
- That there should be freedom of scientific investigation in outer space;
- That outer space is not subject to national appropriation;
- That parties to the treaty retain jurisdiction and control over objects launched into outer space that are listed on their registries, while they are in outer space and that ownership of objects launched into outer space is not affected by their presence in outer space or by their return to Earth;
- That nations be guided by the principle of cooperation and mutual assistance in lunar exploration and use, with due regard to the corresponding interests of other parties to the treaty; and

That international consultations must take place prior to the commencement of an activity that any party has reason to believe would cause potentially harmful interference with activities of other parties.

[17] HR 5063. The text of this Bill is available at: Government Publishing Office http://www.gpo.gov/fdsys/pkg/BILLS-113hr5063ih/pdf/BILLS-113hr5063ih.pdf. For an analysis of the ASTEROIDS Act, see Charles Stotler, "The ASTEROIDS Act and hearing: some observations on international obligations", *The Space Review* (22 September 2014), online: The

introduced in the U. S. House of Representatives. The authors of the bill at that time stated:

Asteroids are excellent potential sources of highly valuable resources and minerals ... that include: platinum group metals such as platinum, osmium, iridium, ruthenium, rhodium, and palladium in addition to nickel, iron and cobalt.[18]

Subsequently, the provisions of this bill were integrated into a new and much broader Space Act of 2015 (formally known as "Spurring Private Aerospace Competitiveness and Entrepreneurship Act of 2015")[19]that was passed by both houses of the U. S. Congress and signed by the president into law on November 25, 2015. A copy of the act is printed in the Appendix to this book.

The new act requires the president, acting through appropriate federal agencies, to facilitate commercial exploration for, and commercial recovery of, space resources by U. S. citizens; discourage government barriers to the development in the United States of economically viable, safe, and stable industries for commercial exploration for and commercial recovery of space resources in manners consistent with the international obligations of the United States; and, promote the right of U. S. citizens to engage in commercial exploration for and commercial recovery of space resources free from harmful interference, in accordance with the international obligations of the United States and subject to authorization and continuing supervision by the federal government.[20]

The act further mandates the president submit to Congress within 180 days of its enactment a report on commercial exploration for and commercial recovery of space resources by U. S. citizens that specifies the authorities necessary to meet the international obligations of the United States, including authorization and continuing supervision by the federal government, and recommendations for the allocation of responsibilities among federal agencies for activities involving commercial exploration for and commercial

---

Space Review, http://www.thespacereview.com/article/2604/1.

[18]"Bipartisan Legislation Promotes Commercial Space Ventures", U.S. Congressman Billy Posey, online: http://posey.house.gov/news/documentprint.aspx?DocumentID=387391. It should be noted that while several space industry representatives and some politicians have been highly optimistic about the significantly profitable space mining, there are some experts who are not so sure about the economic viability of such activities in the near future. Mark Sykes, the Director of the Planetary Science Institute and a co-investigator on the NASA Dawn mission to Vesta and Ceres, has been reported expressing his caution that, "The development of an NEO ISRU infrastructure is beyond the scope of private enterprise........All this basic science and engineering is something beyond the scope of reasonable investment by a commercial entity, because there would be no expectation of return in investment on a reasonable timescales. I expect it would take a couple of decades to get to the point when one could answer the question of whether, with some level infrastructure in place, the marginal cost of processing and returning water from an asteroid would be cheaper than bringing it up from the surface of the Earth." See Mark Strauss, "Congressional Hearing Slams Feasibility Of Commercial Asteroid Mining," 11 September 2014, online at: http://io9.gizmodo.com/congressional-hearing-slams-feasibility-of-commercial-a-1633510688.

[19]*Spurring Private Aerospace Competitiveness and Entrepreneurship Act of 2015* (US), 51 USC Title IV, Pub L 114-90 (Bill HR 2262).

[20]Pub L 114-90 Title IV § 402.

recovery of space resources by U. S. citizens.[21]

While explicitly proclaiming that the enactment of the new act does not amount to an assertion of "sovereignty or sovereign or exclusive rights or jurisdiction over, or the ownership of, any celestial body"[22]on the part of the United States, the act provides nonetheless that:

> *[a] U. S. citizen engaged in commercial recovery of an asteroid resource or a space resource under* [the Act] *shall be entitled to any asteroid resource or space resource obtained, including to possess, own, transport, use, and sell the asteroid resource or space resource obtained in accordance with applicable law, including the international obligations of the United States.*

In essence, the act provides a domestic legal basis for U. S. citizens to engage in commercial exploration and recovery of resources from asteroids, the Moon and other celestial bodies. It is significant to observe that the act makes a distinction between "asteroid resources"—defined as space resources found on or within a single asteroid—and "space resources"—defined as abiotic resources including water and mineral resources in situ in outer space.[23] While the domestic regulatory mechanisms for licensing any such commercial exploration and recovery activities will be determined following the submission of the report of the president to Congress, what is clear from the act is that under U. S. law, any U. S. citizens licensed to carry on commercial exploration and recovery of asteroid resources and space resources will be entitled to ownership of any such resources obtained.

From the U. S. national regulatory perspective, this act is appropriate and necessary to provide statutory and administrative bases for instituting regulating mechanisms and processes. As observed in Chap. 10, the question as to whether or not the new enactment is in conformity with the international obligations of the United States, particularly those arising from the provisions of the 1967 Outer Space Treaty to which the United States is party, is unresolved. A likely response from other spacefaring nations could be to (a) raise this issue at the international level (possibly at the UN COPUOS), and/or (b) enact national laws to provide a regulatory basis and to protect the interests of their citizens who may also wish to commercially explore and exploit natural resources in space, pursuant to both the 1967 Outer Space Treaty and the 1979 Moon Agreement.

The United Kingdom

The United Kingdom is also a State party to the 1967 Outer Space Treaty but not to the Moon Agreement. To a certain extent, the UK's approach towards governmental regulation of space activities follows the comprehensive approach. The Outer Space Act of 1986 governs space activities carried out by organizations established in the UK and by UK nationals wherever they may be.[24] The Secretary of State for Business, Innovation and Skills must license all such activities, a responsibility that is delegated to the British

[21] Pub L 114-90 Title IV § 402.

[22] Pub L 114-90 Title IV§ 403.

[23] Pub L 114-90 Title IV § 401.

[24] *Outer Space Act* (UK) 1986 c. 38.

National Space Centre (BNSC). The application of the Outer Space Act of 1986 has been extended to British Overseas Territories, namely: Guernsey; the Isle of Man; Jersey; and (with modifications) to Gibraltar, Bermuda and the Cayman Islands. The act seeks to ensure compliance with the UK's international obligations in relation to the exploration and use of space, including liability for damage caused by space objects, the registration of objects launched into outer space and the U. N. principles on remote sensing of Earth. All persons and entities to whom the act applies must indemnify the UK government against any claims for damage or loss arising out of licensable activities. This is a mandatory statutory obligation on which no financial limit is set.

A licensee under the UK Outer Space Act must comply with certain terms imposed by the license.[25] These include requirements that the licensee:

- avoid contamination of the space environment and changes to that of Earth;
- avoid interference with the space activities of others;
- dispose of the licensed space object appropriately at the end of the licensed activity and inform the agency of the disposal and termination of the activity;
- avoid any breach of the UK's international obligations;
- inform the agency of any change in the licensed activity and seek approval prior to that change being made;
- obtain insurance against third-party liabilities arising from the licensed activity (both the launch and on-orbit phases), naming the UK government as an additional insured;
- preserve the national security of the UK; and
- permit reasonable access to documents and inspection and testing of equipment and facilities by the agency.

In addition to the Outer Space Act, the Communications Act of 2003[26] also governs the conduct of space activities in the UK inasmuch as the activities concerned involve the use of the radio frequency spectrum. In this regard, it is significant to note that UK laws are in harmony with European Union Directives relevant to communications. In spite of the foregoing, the UK does not have a statute that is exclusively dedicated to the commercial exploitation of the natural resources of outer space by UK nationals or UK entities. Clearly, since any such activities would fall within the broad scope of licensable space activities as defined in the Outer Space Act,[27] it is not farfetched to anticipate that any proposed exploitive activities in space will be regulated under the provisions of the act. So far, the issue has not been put to test, as there have not been any applications from UK nationals or entities for licenses to conduct

[25] *Ibid.* section 5 (2) (e).

[26] *Communications Act* (UK) 2003 c. 21.

[27] *Outer Space Act* (UK), section 1 which provides as follows:

This Act applies to the following activities whether carried out in the United Kingdom or elsewhere:

(a) launching or procuring the launch of a space object;
(b) operating a space object;
(c) any activity in outer space.

exploitive activities on the Moon or other celestial bodies.

The Russian Federation

The Russian Federation, a major space-faring nation, is a State party to the 1967 Outer Space Treaty but not to the 1979 Moon Agreement. In the Russian Federation, space activities are nationally regulated by the federal government under and by virtue of a number of statutes.[28] For our present purposes, the most important Russian statute dealing with space activities is the 1993 Law of the Russian Federation on Space Activities.[29] For the purposes of this law, space activity is defined in Article 2 of the Presidential Decree as any activity immediately connected with operations to explore and use outer space, including the Moon and other celestial bodies, including:

- space research;
- remote sensing of Earth from outer space, including environmental monitoring and meteorology;
- use of navigation, topographical and geodesic satellite systems;
- piloted space missions;
- manufacturing of materials and other products in outer space;
- other kinds of activity performed with the aid of space technology.

In any event, Article 9 of the decree establishes an authorization or licensing procedure for the pursuit of all space activities in the Russian Federation, both for scientific and socioeconomic purposes. The licensing requirements apply to space activities pursued by organizations and citizens of the Russian Federation or those pursued by foreign organizations and citizens but which fall under the jurisdiction of the Russian Federation and involve the manufacture, testing, storage, preparation for launch or launch of space objects, or the control of spaceflights. The licensing and authorization requirements are further elaborated upon in the 1994 Statute of the Russian Federation on Licensing of Space Operations,[30] which defines types forms and periods of validity of licenses, conditions and procedures for their issuance, withholding, suspension or termination and other aspects of licensing. Article 25 of the decree mandates licensees to take out compulsory insurance against damage to the life and health of cosmonauts and personnel on the ground and other objects of space infrastructure, as well as against property damage to third parties.

Article 4 of the decree contains an interesting provision. After listing a

[28] For details, see Sergey P. Malkov and Catherine Doldirina, "Regulation of Space Activities in the Russian Federation," in Ram S Jakhu, ed, *National Regulation of Space Activities* (Heidelberg: Springer, 2010), 315 *et seq.*

[29] Law on Space Activities (Russian Federation) August 20, 1993 as subsequently amended (Decree No. 5663-1 of the Russian House of Soviets). For an unofficial English translation of this statute, see: UN Office of Outer Space Affairs website: http://www.oosa.unvienna.org/oosa/en/SpaceLaw/national/russian_federation/decree_5663-1_E.html.

[30] *Statute on Licensing Space Operations* (Russian Federation) February 2, 1996 (Decree No. 104 of the Russian House of Soviets). Unofficial English translation is available online on UN Office of Outer Space Affairs website: http://www.oosa.unvienna.org/oosa/en/SpaceLaw/national/russian_federation/decree_104_1996E.html.

number of space activities that are specifically prohibited in the Russian Federation, Art. 4(2) continues: "other space activity under the jurisdiction of the Russian Federation, which is prohibited by international treaties of Russian Federation, is not allowed as well." Obviously, space activities that are internationally prohibited by the provisions of international treaties to which the Russian Federation is a State party are also prohibited within the Russian Federation. What remains unclear is whether or not the Russian decree will permit the conduct of space activities that seek to exploit the natural resources of the Moon and other celestial bodies, since there is not much clarity in the provisions of the 1967 Outer Space Treaty about the legality of such activities, and Russia is not a State party to the 1979 Moon Agreement, which arguably provides clearer rules on the matter. Again, it is difficult to predict what the outcome might likely be since, to the knowledge of the present authors, as yet no applications of such nature have been lodged with the relevant Russian authorities.

In light of the foregoing discussion, it is apparent that the Russian Federation does not have a regulatory framework specifically dedicated to govern space activities that aim for the exploitation of the natural resources of outer space. Given that governmental regulation of space activities in the Russian Federation generally follows the comprehensive approach, and having regard to the fact that the definition of space activities is quite broad, it is quite likely that the federal government of the Russian Federation will subject any proposed natural resource exploitation activities in outer space to the licensing and authorization requirements of the Law on Space Activities. Beyond licensing such activities, however, what remains to be seen is the attitude the Russian government might adopt in relation to the international legality or otherwise of such exploitative activities in outer space, particularly those of the U.S. companies.

Australia

Although the Commonwealth of Australia is not a major spacefaring nation, its national space laws are briefly discussed in this section, perhaps as a model worthy of consideration. Australia is a State Party to all of the U. N. Space Law treaties, including the 1979 Moon Agreement, and the government of the Commonwealth regulates space activities that take place in Australian territory as well as those undertaken by Australian citizens outside of Australian territory. Governmental regulation of space activities in Australia occurs under and by virtue of the Space Activities Act of 1998[31]and the Space Activities Regulations of 2001.[32] The act and the regulations are quite comprehensive, requiring different types of licenses and permits for different space operations. For instance, a space license must be obtained before any entity can operate a launch facility or launch vehicle from within the territory of Australia. Launch permits are additionally required for the launching and

[31]*Space Activities Act of 1998* (Cth.) Act No. 123 of 1998 (as subsequently amended).

[32]*Space Activities Regulations* (Cth.) SR 2001 No. 186 (as amended, taking into account amendments up to SR 2004 No. 79).

re-entry of space objects, including those that were originally launched into space from a foreign country. In situations where Australian nationals intend to undertake space activities from places outside the territory of the Commonwealth, they must still obtain overseas launch certificates from the government of Australia before they can engage in those activities.

The provisions of the act and regulations are quite clear and comprehensive. Unfortunately, however, they tend to focus on the regulation of launch and re-entry activities to the neglect of other equally important space operations. Although Australia is a State party to the Moon Agreement, the act and the regulations do not contain specific provisions that address space activities involving the exploitation of the natural resources of the Moon and other celestial bodies. It would seem that the focus of the act and regulations was deliberately placed on launch and re-entry activities given that these are the main types of space activities currently considered important by Australia. Having regard to the comprehensive nature of the provisions of the act and the regulations, and the relative ease of amendment of the regulations in particular, it is fair to argue that they provide a legal basis for the Australian government to regulate any proposed space activities aimed at exploiting the natural resources of the Moon and other celestial bodies. As a State party to the Moon Agreement, the Australian government may confer explicit rights upon its licensees to collect and use the natural resources of the Moon and other celestial bodies as part of its licensing process and in accordance with the provisions of the Moon Agreement.

Canada

Canada does not have a comprehensive national space law in place.[33] This notwithstanding, any person who desires to launch an object into space using rockets classified beyond a specified threshold requires a license issued by the federal minister of transportation under the Canadian Aviation Regulations (CARs).[34] In order to operate any space object(s) using radio frequencies, one requires a radio-communication license issued by the federal minister of industry. If the space activity intended falls within the scope of application of the Remote Sensing Space Systems Act and Regulations, then one also requires a license issued by the federal minister of the Department of Global Affairs (formerly the Department of Foreign Affairs and International Trade).[35] For all other space activities and space applications that fall outside the scope of these areas—such as any proposed resource exploitation activities on the Moon or other celestial bodies—there are no specific regulatory regimes in place as yet. As such, at present, there is no legal or regulatory basis for the Canadian government to license any proposed space

[33] For details, see Ram S. Jakhu, "Regulation of Space Activities in Canada", in Ram S Jakhu, ed, *National Regulation of Space Activities* (Heidelberg: Springer, 2010), 81 et seq.

[34] Canadian Aviation Regulations, Section 602.43 to 604.45, online: http://laws-lois.justice.gc.ca/eng/regulations/SOR-96-433/FullText.html#s-602.43.

[35] Remote Sensing Space Systems Act, S.C. 2005, c. 45. For detailed analysis of the Act, see Ram S. Jakhu, Catherine Doldirina and Yaw Otu Mankata Nyampong, "Review of Canada's Remote Sensing Space Systems Act of 2005," *Annals Of Air And Space Law* (Vol. XXXVII, 2012), 399 et seq.

activities that involve the exploration or exploitation of space's natural resources.

India

India is a major spacefaring nation with ambitious plans for exploration and eventually exploitation of natural resources in space. In addition to its successful mission to the Moon, in 2014, India's Mars Orbiter Mission proved to be the first successful maiden mission of any country, making India the first Asian country to reach Mars with a shoestring budget.[36] India is a State party to the first four U. N. space law treaties and has also signed the 1979 Moon Agreement. For several years India has been attempting to draft its national space law. However, currently there is no specific legislation (except occasionally issued policy guidelines) that regulates space activities,[37] including those involving the exploitation of natural resources in space.

New Legislative Initiatives in Luxembourg and the United Arab Emirates

In the wake of the U.S. Space Act of 2015 and especially the Title 4 legislation with regard to the possibility of U.S. citizens or entities reclaiming space resources, there have now been some responses around the world. In Luxembourg and the United Arab Emirates in particular there have been legislative initiatives that are parallel to the United States that would allow the extraction of resources by industry. In the case of pending Luxembourg legislation that might take effect as early as in 2017, the proposed language would not only create a process for Luxembourg companies, but also would invite companies from other countries to seek to engage in such resource reclamation under the provisions of the Luxembourg law as well. These initiatives will logically spread to other countries until there is a definitive ruling at the international level as to the limitations imposed by the Outer Space Treaty and/or the Moon Agreement.

Conclusions

As demonstrated by the discussion in this section, many spacefaring nations do not have specific legislative and regulatory frameworks established at the national level to address the issues inherent in space activities that target the exploitation of the natural resources in outer space. It appears that even where states have comprehensive national space laws in place, the question of resource exploitation in space is not adequately addressed, probably because there have not been any

[36] Mike Wall, "India's First Mars Probe Makes Historic Red Planet Arrival," September 23, 2014, online: http://www.space.com/27242-india-mars-mission-arrival.html; *India's Mars Orbiter Mission delivers first Science Results, looks at future Challenges,* 11 November 2015, online: http://spaceflight101.com/mom/indias-mars-orbiter-mission-delivers-first-science-results-looks-at-future-challenges/.

[37] For details, see Ranjana Kaul and Ram S. Jakhu, "Regulation of Space Activities in Canada", in Ram S Jakhu, ed, *National Regulation of Space Activities* (Heidelberg: Springer, 2010), 153 *et seq.*

applications for licenses to exploit natural resources in space. The U. S. Space Act of 2015 can be expected to stimulate discussions at national level in several countries for focusing on the development and adoption of some national legal mechanisms to provide a regulatory basis for activities related to exploitation of such natural resources. Whether this leads to new national laws that are essentially parallel to that of the United States or perhaps moves in different directions is still not clear. The consistency of national law with international law and especially the Outer Space Treaty and the Moon Agreement will likely be a key issue of concern and discussion for the next few years to come—and particularly in the context of space mining.

12 Conclusions and the Way Forward

The future of space mining, space exploration, space sciences and space applications is a rich and exciting one. A country-by-country review of various activities involving space exploration and use over the past half-century shows a very broad range of activities that include research, utilization of space applications, exploration by governmental agencies, and now commercial exploration of the Solar System for the purpose of undertaking commercial exploitation of natural resources found in space.

Commercial space is not new. Intelsat and Telesat Canada started commercial communications satellite operations a full half-century ago. Remote sensing and space navigation services are now well established. New commercial activities such as solar power satellite services and on-orbit repairs, retrofitting, and satellite repositioning are next on the horizon. All of these commercial activities, however, are oriented towards providing services directly back to Earth, and all of these enterprises do not involve a true off-world presence. Today, we are entering a truly new age that defines a new milestone for humanity.

We are now contemplating a new tomorrow where humans—and their intelligent robots—work off Earth in space-based enterprises. These new off-world space enterprises will involve not only space mining but also an expanding range of other space industries. These can include operation of off-world habitats, space-based processing and manufacturing, and ultimately what could legitimately be called space colonization and perhaps terraforming of the Moon and Mars.

These off-world enterprises are not seen by today's space entrepreneurs as isolated ventures but ultimately envisioned as being linked together. It has been suggested on a number of occasions that the purpose of space mining is not solely to bring the resources thereby obtained back to Earth but rather to obtain off-world materials for processing in space. Metals and other useful materials obtained in space could then be used to create communications satellites, remote sensing spacecraft, solar power satellites and other finished

© Springer International Publishing Switzerland 2017

R.S. Jakhu et al., *Space Mining and Its Regulation*,
Astronautical Engineering, DOI 10.1007/978-3-319-39246-2_12

products without ever returning to Earth's surface. Ultimately this would allow the creation of full-fledged space manufacturing and processing techniques necessary for the establishment of space colonies and off-world habitats.

There are many writings on how to obtain materials in space so as to allow space manufacturing without the need to obtain any resources from Earth's surface, but sometimes a picture is more powerful than words. Figure 12.1 below provides an illustration of what a solar power satellite system that could provide gigawatts of power to Earth might look like. What is striking about this "vision" is that the entire solar power satellite system would utilize metals, silicas and other materials obtained in space to create this power station in the skies without using any Earth-based natural resources.

This book has sought to explain the global population, urbanization and resource use trends that are pushing human civilization towards a future that involves a space economy and a new reliance on off-world resources. It also warns that, for the future, a new approach to recycling and sustainability is needed since human growth and consumption cannot continue forever unchecked and unmoderated. Nevertheless, space resources can provide a lifeline and new opportunity for the millennia ahead.

Subsequent to this initial analysis, we explored the technological development

Fig. 12.1 A solar power satellite fabricated in the skies from space mined materials (Graphic courtesy of NASA.)

as well as organizational and management innovations that are producing new and lower cost space transportation systems and which, in turn, can make space mining a reality. We also examined the development of small but highly capable remote sensing satellites that can assist in the identification of the most suitable candidates for future space mining operations. This was followed by a discussion of the need for other critical technical developments. These include remote power systems and robotic systems optimized for actual space mining operations.

This is not to say that other capabilities will not be required, but the technology for telecommunications, remote sensing, human habitats, etc., are being developed for other purposes and should be available in the coming decades as and when they are needed. The bottom line is that although the technological developments are demanding and difficult, it is quite likely that these challenges will be overcome. The largest barriers to the future success of space mining are not technical but rather financial, organizational, legal and regulatory.

In the next section of the book, we sought to review space exploration and development activities that have taken place around the world by spacefaring nations. The review showed that so far, the United States has played a leading role in space exploration, noting that the four companies organized to undertake space mining activities (i.e., Deep Space Industries, Planetary Resources, Moon Express, and Shackleton Energy) are all based in the United States. Nevertheless, this global review clearly indicates that other countries have shown a strong interest in space exploration and in future space mining activities that closely parallel those of the United States. In particular the relevant activities of Russia and the former Soviet Union, of Europe, Canada and Australia, Japan, China and India were all explored. These chapters not only summarized the past space exploration missions of the countries mentioned above but also examined their variously expressed interests in space mining and in obtaining economic value from off-world activities.

Finally, in the last part of the book, we explored the current status of the international and national regulatory environment under which difficult and demanding space mining activities may be undertaken. It is clear that current international space law, represented mainly by the 1967 Outer Space Treaty and the 1979 Moon Agreement, will take us only so far. There is the need for a clear global space governance system to provide the basis for orderly exploitation of space's natural resources. National regulatory initiatives such as the U. S. Space Act of 2015 are necessary for national legal and administrative purposes, but could also potentially render the concerned states in violation of their international obligations. Thus, careful implementation of such national laws is important in order to maintain full compliance with relevant international treaties.

It is clear that as the private sector will continue to expand its role in the conduct of space activities, governments will not remain the only key actors in space. However, just as in any other international area of global public interest, the private sector should not be the only stakeholder involved in determining the future potential of the exploitation and use of natural resources in space. The public sector and governments must continue to play a key role. Preferably, there is the need to strike an appropriate balance between the interests of the private sector and those of the public sector and of the governments. Likewise there is also the issue of maintaining a balance between the interests of spacefaring nations and associated industries on the one hand and the interests of non-spacefaring countries on the other.

As discussed in Chap. 11, recent policy and regulatory developments in the United States that relate to the exploration and exploitation of natural resources in space are arousing global interest and concerns. It is quite unlikely that established spacefaring nations, emerging spacefaring nations, non-spacefaring nations and international bodies such as the UN COPUOS will sit idle and do nothing while these ventures unfold. What is likely to happen first of all is a revitalization of discussions within various international forums as to how best to ensure that the exploitation of the natural resources of outer space (particularly led by the private sector) is carried out in an atmosphere that is peaceful, that promotes international cooperation and fosters friendly relations among the States and peoples of the world. International discussions will also likely focus on the environmental and safety aspects of the planned and future resource exploitation activities in space.

It is significant to observe that such discussions have already commenced under the auspices of UN COPUOS. At the 47th Session of the Legal Subcommittee of UN COPUOS held in 2008, a joint statement on the benefits of adherence to the Moon Agreement was presented by some States' parties to the Moon Agreement, namely: Austria, Belgium, Chile, Mexico, the Netherlands, Pakistan and the Philippines.[1] The proponents of the joint statement hoped UN COPUOS would, in the framework of its activities aimed at the development and wider application of outer space law, reflect on some of the benefits of the Moon Agreement. As such, the joint statement, which was based on the experience of the States' parties to the Moon Agreement, was not intended to constitute an authoritative interpretation of the treaties or resolutions mentioned in it, but to emphasize the benefits of certain aspects and considerations of the Moon Agreement and of being a party to it.

The joint statement stated that although the Moon Agreement contains provisions reiterating or elaborating upon the principles contained in the Outer Space Treaty, some of which are directly applicable to the Moon and other celestial bodies, many of its other provisions are unique and have real added value as compared to the other outer space treaties. Some of the provisions unique to the Moon Agreement are of interest for the implementation of projects, activities and missions related to the Moon and celestial bodies in particular since they:

1. clarify or complement principles, procedures and notions found in the

[1] Joint Statement on the benefits of adherence to the Agreement Governing the Activities of States on the Moon and Other Celestial Bodies of 1979 by States Parties to that Agreement; Committee on the Peaceful Uses of Outer Space Legal Subcommittee, Forty-seventh session; UN Doc A/AC.105/C.2/2008/CRP.11 of 2 April 2008.

other outer space treaties that are applicable to the Moon and other celestial bodies; and/or
2. facilitate international scientific cooperation.

After identifying some specific added-value provisions and also discussing the meaning and implications of Article 11 of the Moon Agreement, the joint statement concluded by emphasizing that a better understanding of concepts of international space law and a better description of pertinent concepts and procedures are provided by the Moon Agreement. Above all, the Moon Agreement also represents a mutual commitment by States to find a multilateral framework to facilitate and ensure that the exploitation of the natural resources of celestial bodies is conducted in accordance with general principles of outer space law. The joint statement therefore encouraged States, particularly those considering engaging in forthcoming missions or projects aimed at exploring celestial bodies, to become parties to the Moon Agreement.

Other international efforts relate to deliberations and discussions on the exploration and eventual exploitation of natural resources found in space. For instance, in 2007, 14 of the world's leading space agencies[2] revealed their common vision for a globally coordinated space exploration to the Moon, Mars and beyond in publishing the "Global Exploration Strategy: The Framework for Co-ordination."[3] The document outlined the rationale for society to explore space, defined the then prevailing focus and process of space exploration, the interest in returning to the Moon and exploring Mars, and proposed a framework for the future coordination of global space exploration. A key finding of this framework document was the need to establish a voluntary, non-binding international coordination mechanism, the International Space Exploration Coordination Group (ISECG), through which individual agencies may exchange information regarding interests, objectives, and plans in space exploration with the goal of strengthening both individual exploration programs as well as the collective effort.

On April 10, 2013, the Canadian Space Agency hosted senior representatives from 11 space agencies for a meeting of the ISECG. During the meeting, the group discussed the status of exploration planning, how space exploration could generate benefits for life on Earth and continued work to be reflected in the next edition of the Global Exploration Roadmap. The Global Exploration Roadmap reflects the international effort to define, through continued discussion among space agencies, feasible and sustainable exploration approaches to the Moon, near-Earth asteroids, and Mars. It also takes into account innovative ideas and concepts provided by external stakeholders after the roadmap was first issued in September 2011. The roadmap demonstrates the importance of the International Space Station as

[2] These are: ASI (Italy); BNSC (UK); CNES (France); CNSA (China); CSA (Canada); CSIRO (Australia); DLR (Germany); ESA (European Space Agency); ISRO (India); JAXA (Japan); KARI (Republic of Korea); NASA (USA); NSAU (Ukraine); and, Roscosmos (Russia).

[3] See "Exploring Together: the Global Exploration Strategy", online: European Space Agency http://www.esa.int/Our_Activities/Human_Spaceflight/Exploration/Exploring_together_The_Global_Exploration_Strategy.

a first step and a bridge to exploration of destinations beyond low-Earth orbit. The updated version of the roadmap, published in August 2013, illustrates planned and conceptual near-term missions, which advance human and robotic exploration starting in the Earth-Moon system.[4]

Aside from the discussions in the ISECG, there is no international forum that currently addresses the question of a legal regime for the exploration and exploitation of the natural resources in space. It is therefore recommended that the international community should take some or all of the following steps:

- Invigorate discussions in the ISECG in order to further elaborate its Global Exploration Roadmap in view of recent national regulatory initiatives regarding commercial exploitation of space natural resources;
- States' parties to the Moon Agreement should call a second meeting of the parties (1) to deliberate on the implications of the U. S. Space Act of 2015 with respect to their rights and obligations under the Moon Agreement, (2) to come up a strategy to cope with the situation created by the new U. S. law and (3) to consider how to increase acceptance of, and adopt amendments to, the Moon Agreement, if considered necessary;
- If the required consensus can be achieved, UN COPUOS should initiate consideration of the Moon Agreement, particularly its increased acceptance by States; and,
- A State or a group of States may initiate, thorough the U. N. General Assembly or any other appropriate international forum, a process to seek an Advisory Opinion of the International Court of Justice to clarify the state of current international space law as it concerns commercial exploitation of space's natural resources, particularly in view of recent national regulatory initiatives. This is necessary, as there is a possibility of having divergent views on the interpretation of the Outer Space Treaty as appropriately recognized by the IISL. Moreover, clarification of the applicable international law would be important for avoiding international conflicts and for encouraging the required significant investment in enterprises undertaking exploitation of space's natural resources.

Given the current geopolitical climate, it is quite unlikely that an entirely new treaty governing the exploration and eventual exploitation of the natural resources of the Moon and other celestial bodies will be negotiated and adopted within the next few years. What is likely, however, is that the renewed interest in space exploration and exploitation for natural resources may provide the rationale and impetus for a large number of emerging spacefaring nations and non-spacefaring nations to consider acceding to the Moon Agreement. On the other hand countries that anticipate the possibility of engaging in space mining in the nearer term may take the opposite course of abstaining from ratifying the Moon Agreement. Further they may move to enact domestic legislation that is perhaps parallel to the U. S. legislation adopted in November 2015 entitling exclusive benefits to national public or private entities (corporations). This thus has the

[4] *The Global Exploration Roadmap*, online: NASA http://www.nasa.gov/sites/default/files/files/GER-2013_Small.pdf.

potential of being a divisive issue between those countries and private enterprises that intent on space mining moving in one direction while other countries that view celestial bodies and especially the Moon as part of the global commons moving in the opposite direction. The way forward could be the adherence to Moon Agreement and/or cooperative arrangement for collective venture(s) such as the original INTELSAT or INMARSAT, to which all nations and their respective public and private entities should be invited to participate in order to organise international public private partnerships. The last option would be a kind of an international public-private partnership for exploitation of space's natural resources.

For the moment, leaving aside questions as to whether the 1979 Moon Agreement is the most appropriate international framework for regulating such activities internationally, the fact remains that the development of suitable national legal frameworks to govern such activities is imperative not only for purposes of avoiding conflicts but also as a means of attracting substantial financial investments that are required to fund the exploration and eventual exploitation of the natural resources of space. Appropriate national regulatory frameworks will also be important and necessary for the purposes of (1) national implementation of the international obligation for requiring authorization by, and exercising continuous supervision over, private corporations, and (2) the apportioning of any international liability that the government of a State may be saddled with as a result of damage arising from the exploitative activities conducted in space by a private sector entity. National regulatory frameworks are therefore important, necessary and cannot be overlooked.

The new U. S. Space Act of 2015 represents a key step in terms of national legislation and could perhaps become a model for what other countries may enact in terms of their own efforts to set the stage for future space mining. The U. S. legislation is important not only in terms of seeking to define some national regulatory guidelines related to space mining but also in terms of encouraging the development of new, more efficient space transportation systems, new approaches to liability, and other matters that are ancillary, yet important to the future development of space mining activities. This act will also likely serve as a stimulus to new international discussions about the international legal rules and regulatory controls related to space mining. It will also likely serve to focus on the issue of whether commercial entities may, going forward, be explicitly addressed in international agreements in addition to nation states.

Although the actual mounting of space mining activities may still be some years away, it seems both reasonable and prudent to address the international and national legal, regulatory, financial, and organizational aspects of off-world activities in a proactive way. Waiting until serious problems arise has never worked well in the past. It seems unwise to delay enacting appropriate national and international regulatory controls until after direct conflict between nations or commercial operations have actually occurred. There is the old adage that says: "It makes no sense to bar the barn doors after the horses have been stolen." This applies to precious horses in the barn and to precious metals located in asteroids in space.

Appendix: Excerpts of Key International Space Treaties and Relevant U. S. Law

Treaty on Principles Governing the Activities of States in the Exploration and Use of Outer Space, Including the Moon and Other Celestial Bodies (1967)

(Popularly known as the Outer Space Treaty)

Adoption by the UN General Assembly: 19 December 1966
(Resolution: 2222 (XXI))
Opened for signature: 27 January 1967, London, Moscow, Washington
Entry into force: 10 October 1967
Depositary: Russian Federation, United Kingdom
United States
(Sources: 18 UST 2410; TIAS 6347; 610 UNTS 205)
(As of 4 April 2016, there are 104 ratifications and 25 signatures)

The States Parties to this Treaty,

Inspired by the great prospects opening up before mankind as a result of man's entry into outer space,

Recognizing the common interest of all mankind in the progress of the exploration and use of outer space for peaceful purposes,

Believing that the exploration and use of outer space should be carried on for the benefit of all peoples irrespective of the degree of their economic or scientific development,

Desiring to contribute to broad international co-operation in the scientific as well as the legal aspects of the exploration and use of outer space for peaceful purposes,

Believing that such co-operation will contribute to the development of mutual understanding and to the strengthening of friendly relations between States and peoples,

Recalling resolution 1962 (XVIII), entitled "Declaration of Legal Principles Governing the Activities of States in the Exploration and Use of Outer Space", which was adopted unanimously by the United Nations General Assembly on 13 December 1963,

Recalling resolution 1884 (XVIII), calling upon States to refrain from placing in orbit around the earth any objects carrying nuclear weapons or

© Springer International Publishing Switzerland 2017

R.S. Jakhu et al., *Space Mining and Its Regulation*,
Astronautical Engineering, DOI 10.1007/978-3-319-39246-2

any other kinds of weapons of mass destruction or from installing such weapons on celestial bodies, which was adopted unanimously by the United Nations General Assembly on 17 October 1963,

Taking account of United Nations General Assembly resolution 110 (II) of 3 November 1947, which condemned propaganda designed or likely to provoke or encourage any threat to the peace, breach of the peace or act of aggression, and considering that the aforementioned resolution is applicable to outer space,

Convinced that a Treaty on Principles Governing the Activities of States in the Exploration and Use of Outer Space, including the Moon and Other Celestial Bodies, will further the Purposes and Principles of the Charter of the United Nations,

Have agreed on the following:

Article I

The exploration and use of outer space, including the moon and other celestial bodies, shall be carried out for the benefit and in the interests of all countries, irrespective of their degree of economic or scientific development, and shall be the province of all mankind.

Outer space, including the moon and other celestial bodies, shall be free for exploration and use by all States without discrimination of any kind, on a basis of equality and in accordance with international law, and there shall be free access to all areas of celestial bodies.

There shall be freedom of scientific investigation in outer space, including the moon and other celestial bodies, and States shall facilitate and encourage international co-operation in such investigation.

Article II

Outer space, including the moon and other celestial bodies, is not subject to national appropriation by claim of sovereignty, by means of use or occupation, or by any other means.

Article III

States Parties to the Treaty shall carry on activities in the exploration and use of outer space, including the moon and other celestial bodies, in accordance with international law, including the Charter of the United Nations, in the interest of maintaining international peace and security and promoting international co-operation and understanding.

Article IV

States Parties to the Treaty undertake not to place in orbit around the earth any objects carrying nuclear weapons or any other kinds of weapons of mass destruction, install such weapons on celestial bodies, or station such weapons in outer space in any other manner.

The moon and other celestial bodies shall be used by all States Parties to the Treaty exclusively for peaceful purposes. The establishment of military bases, installations and fortifications, the testing of any type of weapons and the conduct of military manoeuvres on celestial bodies shall be forbidden. The use of military personnel for scientific research or for any other peaceful purposes shall not be prohibited. The use of any equipment or facility necessary for peaceful exploration of the moon and other celestial bodies shall also not be prohibited.

Article V

States Parties to the Treaty shall regard astronauts as envoys of mankind in outer space and shall render to them all possible assistance in the event of accident, distress, or emergency landing on the territory of another State Party or on the high seas. When astronauts make such a landing, they shall be safely and promptly returned to the State of registry of their space vehicle.

In carrying on activities in outer space and on celestial bodies, the astronauts of one State Party shall render all possible assistance to the astronauts of other States Parties.

States Parties to the Treaty shall immediately inform the other States Parties to the Treaty or the Secretary-General of the United Nations of any phenomena they discover in outer space, including the moon and other celestial bodies, which could constitute a danger to the life or health of astronauts.

Article VI

States Parties to the Treaty shall bear international responsibility for national activities in outer space, including the moon and other celestial bodies, whether such activities are carried on by governmental agencies or by non-governmental entities, and for assuring that national activities are carried out in conformity with the provisions set forth in the present Treaty. The activities of non-governmental entities in outer space, including the moon and other celestial bodies, shall require authorization and continuing supervision by the appropriate State Party to the Treaty. When activities are carried on in outer space, including the moon and other celestial bodies, by an international organization, responsibility for compliance with this Treaty shall be borne both by the international organization and by the States Parties to the Treaty participating in such organization.

Article VII

Each State Party to the Treaty that launches or procures the launching of an object into outer space, including the moon and other celestial bodies, and each State Party from whose territory or facility an object is launched, is internationally liable for damage to another State Party to the Treaty or to its natural or juridical persons by such object or its component parts on the Earth, in air space or in outer space, including the moon and other celestial bodies.

Article VIII

A State Party to the Treaty on whose registry an object launched into outer space is carried shall retain jurisdiction and control over such object, and over any personnel thereof, while in outer space or on a celestial body. Ownership of objects launched into outer space, including objects landed or constructed on a celestial body, and of their component parts, is not affected by their presence in outer space or on a celestial body or by their return to the Earth. Such objects or component parts found beyond the limits of the State Party to the Treaty on whose registry they are carried shall be returned to that State Party, which shall, upon request, furnish identifying data prior to their return.

Article IX

In the exploration and use of outer space, including the moon and other celestial bodies, States Parties to the Treaty shall be guided by the principle of co-operation and mutual assistance and shall conduct all their activities in outer space, including the moon and other celestial bodies, with due regard to the corresponding interests of all other States Parties to the Treaty. States

Parties to the Treaty shall pursue studies of outer space, including the moon and other celestial bodies, and conduct exploration of them so as to avoid their harmful contamination and also adverse changes in the environment of the Earth resulting from the introduction of extraterrestrial matter and, where necessary, shall adopt appropriate measures for this purpose. If a State Party to the Treaty has reason to believe that an activity or experiment planned by it or its nationals in outer space, including the moon and other celestial bodies, would cause potentially harmful interference with activities of other States Parties in the peaceful exploration and use of outer space, including the moon and other celestial bodies, it shall undertake appropriate international consultations before proceeding with any such activity or experiment. A State Party to the Treaty which has reason to believe that an activity or experiment planned by another State Party in outer space, including the moon and other celestial bodies, would cause potentially harmful interference with activities in the peaceful exploration and use of outer space, including the moon and other celestial bodies, may request consultation concerning the activity or experiment.

Article XI

In order to promote international co-operation in the peaceful exploration and use of outer space, States Parties to the Treaty conducting activities in outer space, including the moon and other celestial bodies, agree to inform the Secretary-General of the United Nations as well as the public and the international scientific community, to the greatest extent feasible and practicable, of the nature, conduct, locations and results of such activities. On receiving the said information, the Secretary-General of the United Nations should be prepared to disseminate it immediately and effectively.

Article XII

All stations, installations, equipment and space vehicles on the moon and other celestial bodies shall be open to representatives of other States Parties to the Treaty on a basis of reciprocity. Such representatives shall give reasonable advance notice of a projected visit, in order that appropriate consultations may be held and that maximum precautions may be taken to assure safety and to avoid interference with normal operations in the facility to be visited.

Article XVI

Any State Party to the Treaty may give notice of its withdrawal from the Treaty 1 year after its entry into force by written notification to the Depositary Governments. Such withdrawal shall take effect 1 year from the date of receipt of this notification.

Article XVII

This Treaty, of which the English, Russian, French, Spanish and Chinese texts are equally authentic, shall be deposited in the archives of the Depositary Governments. Duly certified copies of this Treaty shall be transmitted by the Depositary Governments to the Governments of the signatory and acceding States.

In witness whereof the undersigned, duly authorised, have signed this Treaty.

Done in triplicate, at the cities of London, Moscow and Washington, the twenty-seventh day of January, one thousand nine hundred and sixty-seven.

Agreement on the Rescue of Astronauts, the Return of Astronauts and the Return of Objects Launched into Outer Space (1968)

Adoption by the UN General Assembly: 19 December 1967
(Resolution: 2345 (XXII))
Opened for signature: 22 April 1968, London, Moscow, Washington
Entry into force: 3 December 1968
Depositary: Russian Federation
United Kingdom
United States
(Sources: 19 UST 7570; TIAS 6599; 672 UNTS 119)
(As of 4 April 2016, there are 94 ratifications, 24 signatures and 2 acceptances of rights and obligations) (Selected Articles)

The Contracting Parties,

Noting the great importance of the Treaty on Principles Governing the Activities of States in the Exploration and Use of Outer Space, including the Moon and Other Celestial Bodies, which calls for the rendering of all possible assistance to astronauts in the event of accident, distress or emergency landing, the prompt and safe return of astronauts, and the return of objects launched into outer space,

Desiring to develop and give further concrete expression to these duties,

Wishing to promote international co-operation in the peaceful exploration and use of outer space,

Prompted by sentiments of humanity,

Have agreed on the following:

Article 1

Each Contracting Party which receives information or discovers that the personnel of a spacecraft have suffered accident or are experiencing conditions of distress or have made an emergency or unintended landing in territory under its jurisdiction or on the high seas or in any other place not under the jurisdiction of any State shall immediately:

(a) notify the launching authority or, if it cannot identify and immediately communicate with the launching authority, immediately make a public announcement by all appropriate means of communication at its disposal;
(b) notify the Secretary-General of the United Nations, who should disseminate the information without delay by all appropriate means of communication at his disposal.

Article 2

If, owing to accident, distress, emergency or unintended landing, the personnel of a spacecraft land in territory under the jurisdiction of a Contracting Party, it shall immediately take all possible steps to rescue them and render them all necessary assistance. It shall inform the launching authority and also the Secretary-General of the United Nations of the steps it is taking and of their progress. If assistance by the launching authority would help to effect a prompt rescue or would contribute substantially to the effectiveness of search and rescue operations, the launching authority shall co-operate with the Contracting Party with a view to the effective conduct of search and rescue operations. Such operations shall be subject to the direction and control of the Contracting Party, which shall act in close and continuing consultation with the launching authority.

Article 3

If information is received or it is discovered that the personnel of a spacecraft have alighted on the high seas or in any other place not under the jurisdiction of any State, those Contracting Parties which are in a position to do so shall, if necessary, extend assistance in search and rescue operations for such personnel to assure their speedy rescue. They shall inform the launching authority and the Secretary-General of the United Nations of the steps they are taking and of their progress.

Article 4

If, owing to accident, distress, emergency or unintended landing, the personnel of a spacecraft land in territory under the jurisdiction of a Contracting Party or have been found on the high seas or in any other place not under the jurisdiction of any State, they shall be safely and promptly returned to representatives of the launching authority.

Article 5

1. Each Contracting Party which receives information or discovers that a space object or its component parts has returned to Earth in territory under its jurisdiction or on the high seas or in any other place not under the jurisdiction of any State, shall notify the launching authority and the Secretary-General of the United Nations.
2. Each Contracting Party having jurisdiction over the territory on which a space object or its component parts has been discovered shall, upon the request of the launching authority and with assistance from that authority if requested, take such steps as it finds practicable to recover the object or component parts.
3. Upon request of the launching authority, objects launched into outer space or their component parts found beyond the territorial limits of the launching authority shall be returned to or held at the disposal of representatives of the launching authority, which shall, upon request, furnish identifying data prior to their return.
4. Notwithstanding paragraphs 2 and 3 of this Article, a Contracting Party which has reason to believe that a space object or its component parts discovered in territory under its jurisdiction, or recovered by it elsewhere, is of a hazardous or deleterious nature may so notify the launching authority, which shall immediately take effective steps, under the direction and control of the said Contracting Party, to eliminate possible danger of harm.
5. Expenses incurred in fulfilling obligations to recover and return a space object or its component parts under paragraphs 2 and 3 of this Article shall be borne by the launching authority.

Article 6

For the purposes of this Agreement, the term "launching authority" shall refer to the State responsible for launching, or, where an international intergovernmental organization is responsible for launching, that organization, provided that organization declares its acceptance of the rights and obligations provided for in this Agreement and a majority of the States members of that organization are Contracting Parties to this Agreement and to the Treaty on Principles Governing the Activities of States in the Exploration and Use of Outer Space, including the Moon and Other Celestial Bodies.

Convention on International Liability for Damage Caused by Space Objects (1972)

Adoption by the UN General Assembly: 29 November 1971
(Resolution: 2777 (XXVI))
Opened for signature: 29 March 1972,
Entry into force: 1 September 1972
Depositary: Russian Federation
United Kingdom
United States
(Sources: 24 UST 2389; TIAS 7762; 961 UNTS 187)
(As of 4 April 2016, there are 92 ratifications, 21 signatures and 3 acceptances of rights and obligations) (Selected Articles)

The States Parties to this Convention,

Recognizing the common interest of all mankind in furthering the exploration and use of outer space for peaceful purposes, Recalling the Treaty on Principles Governing the Activities of States in the Exploration and Use of Outer Space, including the Moon and Other Celestial Bodies,

Taking into consideration that, notwithstanding the precautionary measures to be taken by States and international intergovernmental organizations involved in the launching of space objects, damage may on occasion be caused by such objects,

Recognizing the need to elaborate effective international rules and procedures concerning liability for damage caused by space objects and to ensure, in particular, the prompt payment under the terms of this Convention of a full and equitable measure of compensation to victims of such damage,

Believing that the establishment of such rules and procedures will contribute to the strengthening of international co-operation in the field of the exploration and use of outer space for peaceful purposes,

Have agreed on the following:

Article I

For the purposes of this Convention:

(a) The term "damage" means loss of life, personal injury or other impairment of health; or loss of or damage to property of States or of persons, natural or juridical, or property of international intergovernmental organizations;
(b) The term "launching" includes attempted launching;
(c) The term "launching State" means:
  - a state which launches or procures the launching of a space object;
  - a State from whose territory or facility a space object is launched;
(d) The term "space object" includes component parts of a space object as well as its launch vehicle and parts thereof.

Article II

A launching State shall be absolutely liable to pay compensation for damage caused by its space object on the surface of the earth or to aircraft in flight.

Article III

In the event of damage being caused elsewhere than on the surface of the earth to a space object of one launching State or to persons or property on board such a space object by a space object of another launching State, the latter shall be liable only if the damage is due to its fault or the fault of persons for whom it is responsible.

Article IV

1. In the event of damage being caused elsewhere than on the surface of the earth to a space object of one launching State or to persons or property on board such a space object by a space object of another launching State, and of damage thereby being caused to a third State or to its natural or juridical persons, the first two States shall be jointly and severally liable to the third State, to the extent indicated by the following:
   (a) If the damage has been caused to the third State on the surface of the earth or to aircraft in flight, their liability to the third State shall be absolute;
   (b) If the damage has been caused to a space object of the third State or to persons or property on board that space object elsewhere than on the surface of the earth, their liability to the third State shall be based on the fault of either of the first two States or on the fault of persons for whom either is responsible.
2. In all cases of joint and several liability referred to in paragraph 1 of this Article, the burden of compensation for the damage shall be apportioned between the first two States in accordance with the extent to which they were at fault; if the extent of the fault of each of these States cannot be established, the burden of compensation shall be apportioned equally between them. Such apportionment shall be without prejudice to the right of the third State to seek the entire compensation due under this Convention from any or all of the launching States which are jointly and severally liable.

Article V

1. Whenever two or more States jointly launch a space object, they shall be jointly and severally liable for any damage caused.
2. A launching State which has paid compensation for damage shall have the right to present a claim for indemnification to other participants in the joint launching. The participants in a joint launching may conclude agreements regarding the apportioning among themselves of the financial obligation in respect of which they are jointly and severally liable. Such agreements shall be without prejudice to the right of a State sustaining damage to seek the entire compensation due under this Convention from any or all of the launching States which are jointly and severally liable.
3. A State from whose territory or facility a space object is launched shall be regarded as a participant in a joint launching.

Article VI

1. Subject to the provisions of paragraph 2 of this Article, exoneration from absolute liability shall be granted to the extent that a launching State establishes that the damage has resulted either wholly or partially from gross negligence or from an act or omission done with intent to cause damage on the part of a claimant State or of natural or juridical persons it represents.
2. No exoneration whatever shall be granted in cases where the damage has resulted from activities conducted by a launching State which are not in conformity with international law including, in particular, the Charter of the United Nations and the Treaty on Principles Governing the Activities of States in the Exploration and Use of Outer Space, including the Moon and Other Celestial Bodies.

Article VII

The provisions of this Convention shall not apply to damage caused by a space object of a launching State to:

(a) nationals of that launching State;
(b) foreign nationals during such time as they are participating in the operation of that space object from the time of its launching or at any stage thereafter until its descent, or during such time as they are in the immediate vicinity of a planned launching or recovery area as the result of an invitation by that launching State.

Article VIII

1. A State which suffers damage, or whose natural or juridical persons suffer damage, may present to a launching State a claim for compensation for such damage.
2. If the State of nationality has not presented a claim, another State may, in respect of damage sustained in its territory by any natural or juridical person, present a claim to a launching State.
3. If neither the State of nationality nor the State in whose territory the damage was sustained has presented a claim or notified its intention of presenting a claim, another State may, in respect of damage sustained by its permanent residents, present a claim to a launching State.

Article IX

A claim for compensation for damage shall be presented to a launching State through diplomatic channels. If a State does not maintain diplomatic relations with the launching State concerned, it may request another State to present its claim to that launching State or otherwise represent its interests under this Convention. It may also present its claim through the Secretary-General of the United Nations, provided the claimant State and the launching State are both Members of the United Nations.

Article XII

The compensation which the launching State shall be liable to pay for damage under this Convention shall be determined in accordance with international law and the principles of justice and equity, in order to provide such reparation in respect of the damage as will restore the person, natural or juridical, State or international organization on whose behalf the claim is presented to the condition which would have existed if the damage had not occurred.

Article XIV

If no settlement of a claim is arrived at through diplomatic negotiations as provided for in Article IX, within 1 year from the date on which the claimant State notifies the launching State that it has submitted the documentation of its claim, the parties concerned shall establish a Claims Commission at the request of either party.

Article XXI

If the damage caused by a space object presents a large-scale danger to human life or seriously interferes with the living conditions of the population or the functioning of vital centres, the States Parties, and in particular the launching State, shall examine the possibility of rendering appropriate and rapid assistance to the State which has suffered the damage, when it so requests. However, nothing in this Article shall affect the rights or obligations of the States Parties under this Convention.

Convention on Registration of Objects Launched into Outer Space (1975)

Adoption by the UN General Assembly: 12 November 1974
(Resolution: 3235 (XXIX))
Opened for signature: 14 January 1975, New York
Entry into force: 15 September 1976
Depositary: UN Secretary-General
(Sources: 28 UST 695; TIAS 8480; 1023 UNTS 15)
(As of 4 April 2016, there are 62 ratifications, 4 signatures and 3 acceptances of rights and obligations) (Selected Articles)

The States Parties to this Convention,

Recognizing the common interest of all mankind in furthering the exploration and use of outer space for peaceful purposes,

Recalling that the Treaty on principles governing the activities of States in the exploration and use of outer space, including the moon and other celestial bodies of 27 January 1967 affirms that States shall bear international responsibility for their national activities in outer space and refers to the State on whose registry an object launched into outer space is carried,

Recalling also that the Agreement on the rescue of astronauts, the return of astronauts and the return of objects launched into outer space of 22 April 1968 provides that a launching authority shall, upon request, furnish identifying data prior to the return of an object it has launched into outer space found beyond the territorial limits of the launching authority,

Recalling further that the Convention on international liability for damage caused by space objects of 29 March 1972 establishes international rules and procedures concerning the liability of launching States for damage caused by their space objects,

Desiring, in the light of the Treaty on principles governing the activities of States in the exploration and use of outer space, including the moon and other celestial bodies, to make provision for the national registration by launching States of space objects launched into outer space,

Desiring further that a central register of objects launched into outer space be established and maintained, on a mandatory basis, by the Secretary-General of the United Nations,

Desiring also to provide for States Parties additional means and procedures to assist in the identification of space objects,

Believing that a mandatory system of registering objects launched into outer space would, in particular, assist in their identification and would contribute to the application and development of international law governing the exploration and use of outer space,

Have agreed on the following:

Article I

For the purposes of this Convention:

(a) The term "launching State" means:
- A State which launches or procures the launching of a space object;

- A State from whose territory or facility a space object is launched;

(b) The term "space object" includes component parts of a space object as well as its launch vehicle and parts thereof;

(c) The term "State of registry" means a launching State on whose registry a space object is carried in accordance with article II.

Article II

1. When a space object is launched into earth orbit or beyond, the launching State shall register the space object by means of an entry in an appropriate registry which it shall maintain. Each launching State shall inform the Secretary-General of the United Nations of the establishment of such a registry.
2. Where there are two or more launching States in respect of any such space object, they shall jointly determine which one of them shall register the object in accordance with paragraph 1 of this article, bearing in mind the provisions of article VIII of the Treaty on principles governing the activities of States in the exploration and use of outer space, including the moon and other celestial bodies, and without prejudice to appropriate agreements concluded or to be concluded among the launching States on jurisdiction and control over the space object and over any personnel thereof.
3. The contents of each registry and the conditions under which it is maintained shall be determined by the State of registry concerned.

Article III

1. The Secretary-General of the United Nations shall maintain a Register in which the information furnished in accordance with article IV shall be recorded.
2. There shall be full and open access to the information in this Register.

Article IV

1. Each State of registry shall furnish to the Secretary-General of the United Nations, as soon as practicable, the following information concerning each space object carried on its registry:
   (a) Name of launching State or States;
   (b) An appropriate designator of the space object or its registration number;
   (c) Date and territory or location of launch;
   (d) Basic orbital parameters, including:
      - Nodal period
      - Inclination
      - Apogee
      - Perigee
   (e) General function of the space object.
2. Each State of registry may, from time to time, provide the Secretary-General of the United Nations with additional information concerning a space object carried on its registry.
3. Each State of registry shall notify the Secretary-General of the United Nations, to the greatest extent feasible and as soon as practicable, of space objects concerning which it has previously transmitted information, and which have been but no longer are in earth orbit.

Article V

Whenever a space object launched into earth orbit or beyond is marked with the designator or registration number referred to in article IV, paragraph 1 (b), or both, the State of registry shall notify the Secretary-General of this fact when submitting the information regarding the space object in accordance with article IV. In such case, the Secretary-General of the United Nations shall record this notification in the Register.

Article IX

Any State Party to this Convention may propose amendments to the Convention. Amendments shall enter into force for each State Party to the Convention accepting the amendments upon their acceptance by a majority of the States Parties to the Convention and thereafter for each remaining State Party to the Convention on the date of acceptance by it.

Article XII

The original of this Convention, of which the Arabic, Chinese, English, French, Russian and Spanish texts are equally authentic, shall be deposited with the Secretary-General of the United Nations, who shall send certified copies thereof to all signatory and acceding States.

In witness whereof the undersigned, being duly authorized thereto by their respective Governments, have signed this Convention, opened for signature at New York on the fourteenth day of January one thousand nine hundred and seventy-five.

Agreement Governing the Activities of States on the Moon and Other Celestial Bodies (1979)

(Popularly Known as the Moon Agreement)

Adoption by the UN General Assembly: 5 December 1979
(Resolution: 34/68)
Opened for signature: 18 December 1979, New York
Entry into force: 11 July 1984
Depositary: UN Secretary-General
(Sources: 18 ILM 1434; 1363 UNTS 3)
(As of 4 April 2016, there are 16 ratifications and 4 signatures)

The States Parties to this Agreement,

Noting the achievements of States in the exploration and use of the moon and other celestial bodies,

Recognizing that the moon, as a natural satellite of the earth, has an important role to play in the exploration of outer space,

Determined to promote on the basis of equality the further development of co-operation among States in the exploration and use of the moon and other celestial bodies,

Desiring to prevent the moon from becoming an area of international conflict,

Bearing in mind the benefits which may be derived from the exploitation of the natural resources of the moon and other celestial bodies,

Recalling the Treaty on Principles Governing the Activities of States in the Exploration and Use of Outer Space, including the Moon and Other Celestial Bodies, the Agreement on the Rescue of Astronauts, the Return of Astronauts and the Return of Objects Launched into Outer Space, the Convention on International Liability for Damage Caused by Space Objects, and the Convention on Registration of Objects Launched into Outer Space,

Taking into account the need to define and develop the provisions of these international instruments in relation to the moon and other celestial bodies, having regard to further progress in the exploration and use of outer space,

Have agreed on the following:

Article 1

1. The provisions of this Agreement relating to the moon shall also apply to other celestial bodies within the solar system, other than the earth, except in so far as specific legal norms enter into force with respect to any of these celestial bodies.
2. For the purposes of this Agreement reference to the moon shall include orbits around or other trajectories to or around it.
3. This Agreement does not apply to extraterrestrial materials which reach the surface of the earth by natural means.

Article 2

All activities on the moon, including its exploration and use, shall be carried out in accordance with international law, in particular the Charter of the United Nations, and taking into account the Declaration on Principles of International Law concerning Friendly Relations and Co-operation Among States in accordance with the Charter of the United Nations, adopted by the General Assembly on 24 October 1970, in the interests of maintaining international peace and security and promoting international co-operation and mutual understanding, and with due regard to the corresponding interests of all other States Parties.

Article 3

1. The moon shall be used by all States Parties exclusively for peaceful purposes.
2. Any threat or use of force or any other hostile act or threat of hostile act on the moon is prohibited. It is likewise prohibited to use the moon in order to commit any such act or to engage in any such threat in relation to the earth, the moon, spacecraft, the personnel of spacecraft or man-made space objects.
3. States Parties shall not place in orbit around or other trajectory to or around the moon objects carrying nuclear weapons or any other kinds of weapons of mass destruction or place or use such weapons on or in the moon.
4. The establishment of military bases, installations and fortifications, the testing of any type of weapons and the conduct of military manoeuvres on the moon shall be forbidden. The use of military personnel for scientific research or for any other peaceful purposes shall not be prohibited. The use of any equipment or facility necessary for peaceful exploration and use of the moon shall also not be prohibited.

Article 4

1. The exploration and use of the moon shall be the province of all mankind and shall be carried out for the benefit and in the interests of all countries, irrespective of their degree of economic or scientific development. Due regard shall be paid to the interests of present and future generations as well as to the need to promote higher standards of living and conditions of economic and social progress and development in accordance with the Charter of the United Nations.
2. States Parties shall be guided by the principle of co-operation and mutual assistance in all their activities concerning the exploration and use of the moon. International co-operation in pursuance of this Agreement should be as wide as possible and may take place on a multilateral basis, on a bilateral basis or through international intergovernmental organizations.

Article 5

1. States Parties shall inform the Secretary-General of the United Nations as well as the public and the international scientific community, to the greatest extent feasible and practicable, of their activities concerned with the exploration and use of the moon. Information on the time, purposes, locations, orbital parameters and duration shall be given in respect of each mission to the moon as soon as possible after launching, while information on the results of each mission, including scientific results, shall be furnished upon completion of the mission. In the case of a mission lasting more than 30 days, information on conduct of the mission, including any scientific results, shall be given periodically at 30 days' intervals. For missions lasting more than 6 months, only significant additions to such information need be reported thereafter.
2. If a State Party becomes aware that another State Party plans to operate simultaneously in the same area of or

in the same orbit around or trajectory to or around the moon, it shall promptly inform the other State of the timing of and plans for its own operations.

3. In carrying out activities under this Agreement, States Parties shall promptly inform the Secretary-General, as well as the public and the international scientific community, of any phenomena they discover in outer space, including the moon, which could endanger human life or health, as well as of any indication of organic life.

Article 6

1. There shall be freedom of scientific investigation on the moon by all States Parties without discrimination of any kind, on the basis of equality and in accordance with international law.
2. In carrying out scientific investigations and in furtherance of the provisions of this Agreement, the States Parties shall have the right to collect on and remove from the moon samples of its mineral and other substances. Such samples shall remain at the disposal of those States Parties which caused them to be collected and may be used by them for scientific purposes. States Parties shall have regard to the desirability of making a portion of such samples available to other interested States Parties and the international scientific community for scientific investigation. States Parties may in the course of scientific investigations also use mineral and other substances of the moon in quantities appropriate for the support of their missions.
3. States Parties agree on the desirability of exchanging scientific and other personnel on expeditions to or installations on the moon to the greatest extent feasible and practicable.

Article 7

1. In exploring and using the moon, States Parties shall take measures to prevent the disruption of the existing balance of its environment whether by introducing adverse changes in that environment, by its harmful contamination through the introduction of extra-environmental matter or otherwise. States Parties shall also take measures to avoid harmfully affecting the environment of the earth through the introduction of extraterrestrial matter or otherwise.
2. States Parties shall inform the Secretary-General of the United Nations of the measures being adopted by them in accordance with paragraph 1 of this article and shall also, to the maximum extent feasible, notify him in advance of all placements by them of radio-active materials on the moon and of the purposes of such placements.
3. States Parties shall report to other States Parties and to the Secretary-General concerning areas of the moon having special scientific interest in order that, without prejudice to the rights of other States Parties, consideration may be given to the designation of such areas as international scientific preserves for which special protective arrangements are to be agreed upon in consultation with the competent bodies of the United Nations.

Article 8

1. States Parties may pursue their activities in the exploration and use of the moon anywhere on or below its sur-

face, subject to the provisions of this Agreement.

2. For these purposes States Parties may, in particular:
   (a) Land their space objects on the moon and launch them from the moon;
   (b) Place their personnel, space vehicles, equipment, facilities, stations and installations anywhere on or below the surface of the moon. Personnel, space vehicles, equipment, facilities, stations and installations may move or be moved freely over or below the surface of the moon.
3. Activities of States Parties in accordance with paragraphs 1 and 2 of this article shall not interfere with the activities of other States Parties on the moon. Where such interference may occur, the States Parties concerned shall undertake consultations in accordance with article 15, paragraphs 2 and 3 of this Agreement.

Article 9

1. States Parties may establish manned and unmanned stations on the moon. A State Party establishing a station shall use only that area which is required for the needs of the station and shall immediately inform the Secretary-General of the United Nations of the location and purposes of that station. Subsequently, at annual intervals that State shall likewise inform the Secretary-General whether the station continues in use and whether its purposes have changed.
2. Stations shall be installed in such a manner that they do not impede the free access to all areas of the moon by personnel, vehicles and equipment of other States Parties conducting activities on the moon in accordance with the provisions of this Agreement or of article I of the Treaty on Principles Governing the Activities of States in the Exploration and Use of Outer Space, including the Moon and Other Celestial Bodies.

Article 10

1. States Parties shall adopt all practicable measures to safeguard the life and health of persons on the moon. For this purpose they shall regard any person on the moon as an astronaut within the meaning of article V of the Treaty on Principles Governing the Activities of States in the Exploration and Use of Outer Space, including the Moon and Other Celestial Bodies and as part of the personnel of a spacecraft within the meaning of the Agreement on the Rescue of Astronauts, the Return of Astronauts and the Return of Objects Launched into Outer Space.
2. States Parties shall offer shelter in their stations, installations, vehicles and other facilities to persons in distress on the moon.

Article 11

1. The moon and its natural resources are the common heritage of mankind, which finds its expression in the provisions of this Agreement and in particular in paragraph 5 or this article.
2. The moon is not subject to national appropriation by any claim of sovereignty, by means of use or occupation, or by any other means.
3. Neither the surface nor the subsurface of the moon, nor any part thereof or natural resources in place, shall become property of any State, international intergovernmental or non-governmental organization, national

organization or non-governmental entity or of any natural person. The placement of personnel, space vehicles, equipment, facilities, stations and installations on or below the surface of the moon, including structures connected with its surface or subsurface, shall not create a right of ownership over the surface or the subsurface of the moon or any areas thereof. The foregoing provisions are without prejudice to the international régime referred to in paragraph 5 of this article.

4. States Parties have the right to exploration and use of the moon without discrimination of any kind, on a basis of equality and in accordance with international law and the terms of this Agreement.
5. States Parties to this Agreement hereby undertake to establish an international régime, including appropriate procedures, to govern the exploitation of the natural resources of the moon as such exploitation is about to become feasible. This provision shall be implemented in accordance with article 18 of this Agreement.
6. In order to facilitate the establishment of the international régime referred to in paragraph 5 of this article, States Parties shall inform the Secretary-General of the United Nations as well as the public and the international scientific community, to the greatest extent feasible and practicable, of any natural resources they may discover on the moon.
7. The main purposes of the international régime to be established shall include:
   (a) The orderly and safe development of the natural resources of the moon;
   (b) The rational management of those resources;
   (c) The expansion of opportunities in the use of those resources;
   (d) An equitable sharing by all States Parties in the benefits derived from those resources, whereby the interests and needs of the developing countries, as well as the efforts of those countries which have contributed either directly or indirectly to the exploration of the moon, shall be given special consideration.
8. All the activities with respect to the natural resources of the moon shall be carried out in a manner compatible with the purposes specified in paragraph 7 of this article and the provisions of article 6, paragraph 2, of this Agreement.

Article 12

1. States Parties shall retain jurisdiction and control over their personnel, vehicles, equipment, facilities, stations and installations on the moon. The ownership of space vehicles, equipment, facilities, stations and installations shall not be affected by their presence on the moon.
2. Vehicles, installations and equipment or their component parts found in places other than their intended location shall be dealt with in accordance with article 5 of the Agreement on Rescue of Astronauts, the Return of Astronauts and the Return of Objects Launched into Outer Space.
3. In the event of an emergency involving a threat to human life, States Parties may use the equipment, vehicles, installations, facilities or supplies of other States Parties on the moon. Prompt notification of such

use shall be made to the Secretary-General of the United Nations or the State Party concerned.

Article 13

A State Party which learns of the crash landing, forced landing or other unintended landing on the moon of a space object, or its component parts, that were not launched by it, shall promptly inform the launching State Party and the Secretary-General of the United Nations.

Article 14

1. States Parties to this Agreement shall bear international responsibility for national activities on the moon, whether such activities are carried on by governmental agencies or by non-governmental entities, and for assuring that national activities are carried out in conformity with the provisions set forth in this Agreement. States Parties shall ensure that non-governmental entities under their jurisdiction shall engage in activities on the moon only under the authority and continuing supervision of the appropriate State Party.
2. States Parties recognize that detailed arrangements concerning liability for damage caused on the moon, in addition to the provisions of the Treaty on Principles Governing the Activities of States in the Exploration and Use of Outer Space, including the Moon and Other Celestial Bodies and the Convention on International Liability for Damage Caused by Space Objects, may become necessary as a result of more extensive activities on the moon. Any such arrangements shall be elaborated in accordance with the procedure provided for in article 18 of this Agreement.

Article 15

1. Each State Party may assure itself that the activities of other States Parties in the exploration and use of the moon are compatible with the provisions of this Agreement. To this end, all space vehicles, equipment, facilities, stations and installations on the moon shall be open to other States Parties. Such States Parties shall give reasonable advance notice of a projected visit, in order that appropriate consultations may be held and that maximum precautions may be taken to assure safety and to avoid interference with normal operations in the facility to be visited. In pursuance of this article, any State Party may act on its own behalf or with the full or partial assistance of any other State Party or through appropriate international procedures within the framework of the United Nations and in accordance with the Charter.
2. A State Party which has reason to believe that another State Party is not fulfilling the obligations incumbent upon it pursuant to this Agreement or that another State Party is interfering with the rights which the former State has under this Agreement may request consultations with that State Party. A State Party receiving such a request shall enter into such consultations without delay. Any other State Party which requests to do so shall be entitled to take part in the consultations. Each State Party participating in such consultations shall seek a mutually acceptable resolution of any controversy and shall bear in mind the rights and interests of all States Parties. The Secretary-General of the United Nations shall be informed of the results of the con-

sultations and shall transmit the information received to all States Parties concerned.

3. If the consultations do not lead to a mutually acceptable settlement which has due regard for the rights and interests of all States Parties, the parties concerned shall take all measures to settle the dispute by other peaceful means of their choice appropriate to the circumstances and the nature of the dispute. If difficulties arise in connection with the opening of consultations or if consultations do not lead to a mutually acceptable settlement, any State Party may seek the assistance of the Secretary-General, without seeking the consent of any other State Party concerned, in order to resolve the controversy. A State Party which does not maintain diplomatic relations with another State Party concerned shall participate in such consultations, at its choice, either itself or through another State Party or the Secretary-General as intermediary.

Article 16

With the exception of articles 17 to 21, references in this Agreement to States shall be deemed to apply to any international intergovernmental organization which conducts space activities if the organization declares its acceptance of the rights and obligations provided for in this Agreement and if a majority of the States members of the organization are States Parties to this Agreement and to the Treaty on Principles Governing the Activities of States in the Exploration and Use of Outer Space, including the Moon and Other Celestial Bodies. States members of any such organization which are States Parties to this Agreement shall take all appropriate steps to ensure that the organization makes a declaration in accordance with the foregoing.

Article 17

Any State Party to this Agreement may propose amendments to the Agreement. Amendments shall enter into force for each State Party to the Agreement accepting the amendments upon their acceptance by a majority of the States Parties to the Agreement and thereafter for each remaining State Party to the Agreement on the date of acceptance by it.

Article 18

Ten years after the entry into force of this Agreement, the question of the review of the Agreement shall be included in the provisional agenda of the General Assembly of the United Nations in order to consider, in the light of past application of the Agreement, whether it requires revision. However, at any time after the Agreement has been in force for 5 years, the Secretary-General of the United Nations, as depository, shall, at the request of one third of the States Parties to the Agreement and with the concurrence of the majority of the States Parties, convene a conference of the States Parties to review this Agreement. A review conference shall also consider the question of the implementation of the provisions of article 11, paragraph 5, on the basis of the principle referred to in paragraph 1 of that article and taking into account in particular any relevant technological developments.

Article 19

1. This Agreement shall be open for signature by all States at United Nations Headquarters in New York.

2. This Agreement shall be subject to ratification by signatory States. Any State which does not sign this Agreement before its entry into force in accordance with paragraph 3 of this article may accede to it at any time. Instruments of ratification or accession shall be deposited with the Secretary-General of the United Nations.
3. This Agreement shall enter into force on the thirtieth day following the date of deposit of the fifth instrument of ratification.
4. For each State depositing its instrument of ratification or accession after the entry into force of this Agreement, it shall enter into force on the thirtieth day following the date of deposit of any such instrument.
5. The Secretary-General shall promptly inform all signatory and acceding States of the date of each signature, the date of deposit of each instrument of ratification or accession to this Agreement, the date of its entry into force and other notices.

Article 20

Any State Party to this Agreement may give notice of its withdrawal from the Agreement 1 year after its entry into force by written notification to the Secretary-General of the United Nations. Such withdrawal shall take effect 1 year from the date of receipt of this notification.

Article 21

The original of this Agreement, of which the Arabic, Chinese, English, French, Russian and Spanish texts are equally authentic, shall be deposited with the Secretary-General of the United Nations, who shall end certified copies thereof to all signatory and acceding States.

In witness whereof the undersigned, being duly authorized thereto by their respective Governments, have signed this Agreement, opened for signature at New York on December 18, 1979.

The United States: Space Resource Exploration and Utilization Act of 2015

Note this act is Title IV of the Space Act of 2015

One Hundred Fourteenth Congress of the United States of America

At the First Session

Begun and held at the City of Washington on Tuesday, the sixth day of January, two thousand and fifteen

An Act

To facilitate a pro-growth environment for the developing commercial space industry by encouraging private sector investment and creating more stable and predictable regulatory conditions, and for other purposes.

Be it enacted by the Senate and House of Representatives of the United States of America in Congress assembled,

Title IV: Space Resource Exploration and Utilization

SEC. 401. Short Title

This title may be cited as the "Space Resource Exploration and Utilization Act of 2015".

SEC. 402. Title 51 Amendment

(a) IN GENERAL.—Subtitle V is amended by adding at the end the following:

"Chapter 513: Space Resource Commercial Exploration and Utilization

"§ 51301. Definitions

"In this chapter:

"(1) ASTEROID RESOURCE.—The term 'asteroid resource' means a space resource found on or within a single asteroid.

"(2) SPACE RESOURCE.—

"(A) IN GENERAL.—The term 'space resource' means an abiotic resource in situ in outer space.

"(B) INCLUSIONS.—The term 'space resource' includes water and minerals.

"(3) UNITED STATES CITIZEN.—The term 'United States citizen' has the meaning given the term 'citizen of the United States' in section 50902.

"§ 51302. Commercial Exploration and Commercial Recovery

"(a) IN GENERAL.—The President, acting through appropriate Federal agencies, shall—

"(1) facilitate commercial exploration for and commercial recovery of space resources by United States citizens;

"(2) discourage government barriers to the development in the United States of economically viable, safe, and stable industries for commercial exploration for and commercial recovery of space resources in manners consistent with the international obligations of the United States; and

"(3) promote the right of United States citizens to engage in commercial exploration for and commercial recovery of space resources free from harmful interference, in accordance with the international obligations of the United States and subject to authorization and continuing supervision by the Federal Government.

"(b) REPORT.—Not later than 180 days after the date of enactment of this section, the President shall submit to Congress a report on commercial exploration for and commercial recovery of space resources by United States citizens that specifies—

"(1) the authorities necessary to meet the international obligations of the United States, including authorization and continuing supervision by the Federal Government; and

"(2) recommendations for the allocation of responsibilities among Federal agencies for the activities described in paragraph (1).

"§ 51303. Asteroid Resource and Space Resource Rights

"A United States citizen engaged in commercial recovery of an asteroid resource or a space resource under this chapter shall be entitled to any asteroid resource or space resource obtained, including to possess, own, transport, use, and sell the asteroid resource or space resource obtained in accordance with applicable law, including the international obligations of the United States."

(b) TABLE OF CHAPTERS.—The table of chapters for title 51 is amended by adding at the end of the items for subtitle V the following:

"513. Space resource commercial exploration and utilization ....0.51301".

SEC. 403. Disclaimer of Extraterritorial Sovereignty

It is the sense of Congress that by the enactment of this Act, the United States does not thereby assert sovereignty or sovereign or exclusive rights or jurisdiction over, or the ownership of, any celestial body.

Speaker of the House of Representatives.

Vice President of the United States and

President of the Senate.

Glossary

Amor orbit NEO orbit that does not cross Earth's orbit and is always outside Earth's orbit yet within 0.03 AU.

Apollo orbit NEO orbit that crosses Earth's orbit twice a year with an orbital size that is larger than Earth's orbit.

Assistance Convention Convention on Assistance in the Case of a Nuclear Accident or Radiological Emergency, adopted on February 26, 1987.

Aten orbit NEO orbit that crosses Earth orbit twice a year with an orbital size that is smaller than Earth orbit.

CATALYST NASA's developmental partnership with industry partners carried out under NASA's Space Act Agreements (SAAs). The acronym stands for Lunar CArgo Transportation And Landing bY Soft Touchdown (Lunar CATALYST).

COPUOS (Also UN COPUOS) Committee on the Peaceful Uses of Outer Space of the United Nations.

DSI Deep Space Industries

EPOXI EPOXI combines two targets: the Deep Impact Extended Investigation (DIXI), and the Extrasolar Planet Observation and Characterization (EPOCh). The *Deep Impact* spacecraft will conduct both the DIXI and the Extrasolar Planet Observation and Characterization missions during the cruise phase to Comet 103P/Hartley. The spacecraft was also used as a test platform for a delay-tolerant networking transmission while at a distance of 20 million miles from Earth.

HST The Hubble Space Telescope

InSight This acronym for the U. S. Mars exploratory land mission stands for "Interior exploration using Seismic Investigations, Geodesy and Heat Transport."

IAASS International Association for the Advancement of Space Safety. This is headquartered to Norwijk, The Netherlands near the European Space Agency ESTEC facility.

IEOs NEOs with an inner Earth orbit. NEO orbit that does not cross Earth's orbit and is always inside Earth's orbit yet within 0.03 AU.

© Springer International Publishing Switzerland 2017
R.S. Jakhu et al., *Space Mining and Its Regulation*,
Astronautical Engineering, DOI 10.1007/978-3-319-39246-2

ITU The International Telecommunication Union.

James Webb Space Telescope The new high performance telescope that is intended to replace the Hubble Space Telescope. This is a NASA and ESA joint project that is scheduled for launch in 2018.

LADEE The NASA Lunar Atmosphere and Dust Environment Explorer (LADEE) mission.

LCROSS The Lunar CRater Observing and Sensing Satellite (LCROSS) mission.

Liability Convention The Convention on International Liability for Damage Caused by Space Objects, adopted by the U. N. General Assembly in its resolution 2777 (XXVI), opened for signature on March 29, 1972, and entered into force on September 1, 1972.

LP The NASA Lunar Prospector (LP) mission.

LRO The NASA Lunar Reconnaissance Orbiter (LRO) mission.

LTSSA UN COPUOS Working Group on the Long Term Sustainability of Space Activities.

MAVEN The NASA Mars Atmosphere and Volatile EvolutioN (MAVEN) mission in November 2013 to explore Mars's upper atmosphere, ionosphere and interactions with the Sun and solar wind.

MER The NASA Mars Exploration Program.

MERLIN This acronym is short for the Moon Express Robotics Lab for Innovation and is located in Mountain View, California, at the Moon Express headquarters.

Moon Agreement (Treaty) The Agreement Governing the Activities of States on the Moon and Other Celestial Bodies, adopted by the U. N. General Assembly in its resolution 34.68, opened for signature on December 18, 1979, entered into force on July 11, 1984.

Moon Express A privately funded U. S. commercial space company created to develop the resources of the Moon, aims to "send a series of robotic spacecraft to the Moon for ongoing exploration and commercial development": http://www.moonexpress.com/#missions.

MSL The acronym for the NASA Mars Science Laboratory spacecraft. It includes the Mars rover Curiosity that was launched on November 26, 2011, and landed successfully on August 6, 2012. The mission of Curiosity was to investigate "whether conditions have been favorable for microbial life and for preserving clues in the rocks about possible past life."

NEA Near Earth asteroid.

Near Earth Orbit Object Survey Act Section 321 of the NASA Authorization Act of 2005 (Public Law No. 109-155), also known as the George E. Brown, Jr. Near-Earth Object Survey Act.

NEO Near Earth object.

NEOCAM The Near Earth Orbit Camera spacecraft. This is a proposed mission by NASA to assist in the location and orbit determination for NEOs.

NEOWISE The Near Earth Orbit Wide-range Infra-red Surveyor Explorer.

New Horizons This is the Pluto and Kuiper Belt exploratory mission.

Notification Convention Convention on Early Notification of a Nuclear Accident, adopted on October 27, 1986.

OSIRIS-REx The Origins Spectral Interpretation Resource Identification Security—Regolith Explorer (OSIRIS-REx) mission that will begin in late 2016. The OSIRIS-REx spacecraft will "travel to a near-Earth asteroid, called Bennu (formerly 1999 RQ36). The objective is to bring back to Earth at least a 2.1-oz sample for study.

Outer Space Treaty (OST) The Treaty on Principles Governing the Activities of States in the Exploration and Use of Outer Space, including the Moon and Other Celestial Bodies, adopted by the U. N. General Assembly in its Resolution 2222 (XXI), opened for signature on January 27, 1967, entered into force on October 10, 1967.

PHA Potentially hazardous asteroid.

PRI Planetary Resources.

Registration Convention The Convention on Registration of Objects Launched into Outer Space, adopted by the U. N. General Assembly in its Resolution 3235 (XXIX), opened for signature on January 14, 1975, entered into force on September 15, 1976.

Rescue and Return Agreement The Agreement on the Rescue of Astronauts, the Return of Astronauts and the Return of Objects Launched into Outer Space (the "Rescue Agreement"), adopted by the General Assembly in its resolution 2345 (XXII), opened for signature on April 22, 1968, entered into force on December 3, 1968.

ROSETTA A mission of the European Space Agency (ESA) to land on a comet and determine its composition by means of the Philae lander.

SSA NASA Space Act Agreement. Also, space situational awareness.

Stardust The first NASA mission, launched in 1999, to obtain a sample from Comet Wild-2.

Stardust NeXT The extended mission of Stardust, called Stardust-New Exploration of Tempel. This is designed to accomplish in 2011 a close flyby of Comet Tempel-1.

SWAS Submillimeter Wave Astronomy Satellite.

U. N. United Nations, also known as the United Nations Organization, or UNO.

WISE Wide-range Infra-red Surveyor Explorer. Also see NEOWISE.

WFS World Future Society.

Index

© Springer International Publishing Switzerland 2017
R.S. Jakhu et al., *Space Mining and Its Regulation*, Astronautical Engineering, DOI 10.1007/978-3-319-39246-2

The manufacturer's authorised representative in the EU is Springer Nature Customer Service Centre GmbH, Europaplatz 3, 69115 Heidelberg, Germany. If you have any concerns regarding our products, please contact ProductSafety@springernature.com

Printed and bound by CPI Group (UK) Ltd, Croydon, CR0 4YY

30/09/2026

02225587-0012